Analytical Chemistry and Microchemistry

nova
science publishers
www.novapublishers.com

Analytical Chemistry and Microchemistry

Modern Approaches in Fluid Chromatography: Impact and Applications
Rohit Bhatia, PhD and Bhupinder Kumar, PhD (Editors)
2023. ISBN: 979-8-88697-856-8 (Hardcover)
2023. ISBN: 979-8-89113-022-7 (eBook)

Chromatographic Methods and Research
Raven W. Obrien (Edtor)
2023. ISBN: 979-8-89113-193-4 (Hardcover)
2023. ISBN: 979-8-89113-203-0 (eBook)

A Guide to Near-Infrared Spectroscopy
Juan Francisco García Martín, PhD (Editor)
2023. ISBN: 979-8-88697-775-2 (Hardcover)
2023. ISBN: 979-8-88697-968-8 (eBook)

Airborne and Biological Monitoring to Assess Occupational Exposure to Isocyanates
Jimmy Hu, PhD
2023. ISBN: 979-8-89113-010-4 (Softcover)
2023. ISBN: 979-8-89113-099-9 (eBook)

Chemometrics: Advances in Applications and Research
Larry D. Crenshaw (Editor)
2022. ISBN: 979-8-88697-274-0 (Hardcover)
2022. ISBN: 979-8-88697-311-2 (eBook)

More information about this series can be found at
https://novapublishers.com/product-category/series/analytical-chemistry-and-microchemistry/

Analytical Chemistry and Microchemistry

nova science publishers

www.novapublishers.com

Analytical Chemistry and Microchemistry

Modern Approaches in Fluid Chromatography: Impact and Applications
Rohit Bhatia, PhD and Bhupinder Kumar, PhD (Editors)
2023. ISBN: 979-8-88697-856-8 (Hardcover)
2023. ISBN: 979-8-89113-022-7 (eBook)

Chromatographic Methods and Research
Raven W. Obrien (Edtor)
2023. ISBN: 979-8-89113-193-4 (Hardcover)
2023. ISBN: 979-8-89113-203-0 (eBook)

A Guide to Near-Infrared Spectroscopy
Juan Francisco García Martín, PhD (Editor)
2023. ISBN: 979-8-88697-775-2 (Hardcover)
2023. ISBN: 979-8-88697-968-8 (eBook)

Airborne and Biological Monitoring to Assess Occupational Exposure to Isocyanates
Jimmy Hu, PhD
2023. ISBN: 979-8-89113-010-4 (Softcover)
2023. ISBN: 979-8-89113-099-9 (eBook)

Chemometrics: Advances in Applications and Research
Larry D. Crenshaw (Editor)
2022. ISBN: 979-8-88697-274-0 (Hardcover)
2022. ISBN: 979-8-88697-311-2 (eBook)

More information about this series can be found at
https://novapublishers.com/product-category/series/analytical-chemistry-and-microchemistry/

Rohit Bhatia, PhD
and Bhupinder Kumar, PhD
Editors

Modern Approaches in Fluid Chromatography

Impact and Applications

nova
science publishers
www.novapublishers.com

NOTICE TO THE READER

Library of Congress Cataloging-in-Publication Data

ISBN: 979-8-88697-856-8

Published by Nova Science Publishers, Inc. † New York

Contents

Preface

Liquid chromatography is one of most predominantly used chromatographic technique. It is broad classification of chromatography which include different type of methods based on different types of mechanisms of separation such as HPLC, Capillary electrophoresis, Ion Exchange, TLC, HPTLC, Paper Chromatography, etc. The efficiency of separation of components depends up on number of factors which need to be taken care while performing chromatography. Technological advancements in fluid chromatography are boosting the demand of these techniques in pharmaceutical industries, research and development, food industries, bio-analysis, forensics and to address issues regarding safety and authenticity of drugs. Time-to-time advancements in the stationary phases to instrumentation and mobile phases to method development have provided success in these areas. With advancements, some problems still remain persistent in such techniques. The various studies are reported separately to counter these persistent problems with various modifications with reproducible results.

This edition of book covers the basic concepts, recent advancements and significant applications of various fluid chromatographic techniques under one cover. This book will provide an upshot in single spot solution to find latest information on advancements in fluid chromatography such as choice of solvents, choice of columns, choice of detectors and other chromatographic conditions. This is a good text for a lecture course on advanced fluid chromatography aimed at advanced undergraduates or, indeed, anyone who finds that they would like to strengthen their understanding of separation science without necessarily going on to expert status. This is also a troubleshooting record to solve complex analytical problems such as sensitivity, difficulty in detection and reproducibility. Following are the key highlights of the contents of this book: i) Introduction to the need and basic principles of fluid chromatography; ii) Detailed description of advanced fluid chromatographic techniques; iii) Detailed instrumentations of various fluid chromatographic techniques; and iv) Recent applications of fluid chromatography in various analytical fields.

Chapter 1

Fluid Chromatography: An Introduction

Rohit Bhatia[1,*]
and Bhupinder Kumar[2]

[1]Department of Pharmaceutical Analysis, ISF College of Pharmacy,
Ghal Kalan, Moga, Punjab, India
[2]Department of Pharmaceutical Sciences, HNB Garhwal University, Chauras Campus,
Srinagar, Garhwal, Uttarakhand, India

Abstract

Chromatography is an excellent technique for qualitative and quantitative analysis. It is used for the separation as well as purification of small molecules, xenobiotics, large complex mixtures, proteins as well as nucleic acids. Fluid (Liquid) chromatography is the most widely utilized type of chromatography. Technological advancements in fluid chromatography are boosting the demand for these techniques in pharmaceutical industries, research and development, food industries, bio-analysis, forensics and addressing issues regarding the safety and authenticity of drugs. On the basis of type of analyte and mode of operation, it can be categorized into various types such as high-performance liquid chromatography (HPLC), high-performance thin layer chromatography (HPTLC), supercritical fluid chromatography (SCFC or SFC), size exclusion chromatography, capillary electrophoresis, etc. Recent advancements have helped in the upgradation of these techniques in achieving high resolution of complex mixtures and reducing time and cost per sample. Hyphenation of these traditional techniques with spectral analytical techniques has further provided key advancements in separation along with analysis. In this book, various fluid chromatographic techniques along with their

* Corresponding Author's Email: bhatiarohit5678@gmail.com.

In: Modern Approaches in Fluid Chromatography
Editors: Rohit Bhatia and Bhupinder Kumar
ISBN: 979-8-88697-856-8

hyphenation, advancements and applications in various fields are compiled. This book will provide key insights on instrumentation, their advancements and achievements in various fields such as chemical, biochemical, pharmaceutical, agriculture, etc.

Keywords: chromatography, SCFC, capillary electrophoresis, 2D liquid chromatography, HPTLC, HPLC

1. Introduction

Chromatography is a separation and purification analytical technique used for both quantitative as well as qualitative purposes. It is not only used for the separation and purification of small organic molecules but also various biomolecules such as nucleic acids, proteins, DNA, RNA or any other complex mixtures that can be individualized for further analysis and other purposes [1, 2]. Chromatography can be used for both analytical and preparative purposes based on the requirement. The different parts of chromatography are the stationary phase, the mobile phase and the molecules which need to be separated or analysed [3]. The components of the mixture separate based on their migration rate toward the stationary phase in particular conditions. Each particle has a different rate of migration under a particular set of conditions which helps in the resolution of complex mixtures via chromatography [4, 5]. Majorly, based on the type of mobile phase, it can be categorized into two classes i.e., liquid chromatography and gas chromatography [6].

Liquid chromatography is one of the most predominantly used chromatographic techniques in which the mobile phase is liquid. It is a broad classification of chromatography which include different type of methods based on different types of mechanisms of separation [7]. The liquid chromatography is carried out through a stationary phase either held in a column (HPLC, Capillary electrophoresis, Ion Exchange, etc.) or against a plane (TLC, HPTLC, Paper Chromatography etc.). Based on physiochemical interactions of solute with stationary and mobile phases different modes of separation include adsorption, partitioning, affinity, size exclusion [8], ion exchange [9], hydrophobic and charged interactions. The efficiency of the separation of components depends on a number of factors which need to be taken care of while performing chromatography. Other than a stationary and mobile phase, factors such as elution mode (isocratic or gradients), flow rate, column length, temperature, etc. are sought to play a key role in the separation and resolution of the mixture [10, 11].

Technological advancements in fluid chromatography are boosting the demand for these techniques in pharmaceutical industries, research and development, food industries, bio-analysis, forensics and addressing issues regarding the safety and authenticity of drugs. This edition of the book covers the basic concepts, recent advancements and significant applications of various fluid chromatographic techniques under one cover. This book provides an upshot in single spot solution to find the latest information on advancements in fluid chromatography such as choice of solvents, choice of columns, choice of detectors and other chromatographic conditions. This is a good text for a lecture course on advanced fluid chromatography aimed at advanced undergraduates or, indeed, anyone who finds that they would like to strengthen their understanding of separation science without necessarily going on to expert status. This is also a troubleshooting record to solve complex analytical problems such as sensitivity, difficulty in detection and reproducibility. In this book, we have provided information regarding various fluid-based chromatographic methods which include Ultra high-performance liquid chromatography and its hyphenation with mass spectrometry (UHPLC-MS) [12], hydrophilic interaction liquid chromatography (HILIC), supercritical fluid chromatography (SCFC or SFC), High-performance thin layer chromatography (HPTLC), capillary electrophoresis and their advancements such as two-dimensional (2D) liquid chromatography [13].

Following are the key highlights of the contents of this book:

- Introduction to the need and basic principles of fluid chromatography
- Detailed description of advanced fluid chromatographic techniques
- Detailed instrumentations of various fluid chromatographic techniques
- Recent applications of fluid chromatography in various analytical fields

Conclusion

The subject of the proposed book is to make readers familiar with and aware of various kinds of fluid chromatographic techniques including their instrumentation and recent advancements. Beyond this, we also summarized the recently reported works in various analytical fields which will highlight the various optimized conditions and significant outcomes. Therefore, this book will be helpful to the undergraduate and post-graduate students to enhance their knowledge and also to select the suitable technology for their research work.

References

[1] Coskun O., *Separation techniques: chromatography*, Northern clinics of Istanbul, 3 (2016) 156.
[2] Poole C. F., *The essence of chromatography*, Elsevier, 2003.
[3] Scott R. P., *Principles and practice of chromatography*, Chrom-ed book series, 1 (2003).
[4] Miller, J. M. *Chromatography: concepts and contrasts*, John Wiley & Sons, 2005.
[5] Cazes J., R. P. Scott, *Chromatography theory*, CRC press, 2002.
[6] D.B. Hibbert, Experimental design in chromatography: a tutorial review, *Journal of chromatography B*, 910 (2012) 2-13.
[7] Fanali S., P. R. Haddad, C. F. Poole, M.-L. Riekkola, *Liquid chromatography: applications*, Elsevier, 2017.
[8] Barth H. G., C. Jackson, B. E. Boyes, *Size exclusion chromatography, Analytical chemistry*, 66 (1994) 595-620.
[9] Weiss J., *Handbook of Ion Chromatography*, 3 Volume Set, John Wiley & Sons, 2016.
[10] Scott R. P., *Techniques and practice of chromatography*, CRC Press, 2020.
[11] Poole C. F., S.A. Schuette, *Contemporary practice of chromatography*, Elsevier, 2012.
[12] Niessen W. M., *Liquid chromatography-mass spectrometry*, CRC press, 2006.
[13] Dugo, P. F. Cacciola, T. Kumm, G. Dugo, L. Mondello, Comprehensive multidimensional liquid chromatography: theory and applications, *Journal of Chromatography* A, 1184 (2008) 353-368.

Chapter 2

Ultra High-Performance Liquid Chromatography Coupled Mass with Spectrometry (UHPLC-MS): Basic Principles and Applications

Divya Shikha[1,2,*]
and Amandeep Singh[3,†]
[1]Department of Pharmaceutical Analysis, ISF College of Pharmacy, Ghal Kalan, Punjab, India
[2]Department of Pharmacy, Usha Martin University Angara, Ranchi, Jharkhand, India
[3]Department of Pharmaceutics, ISF College of Pharmacy, Ghal Kalan, Punjab, India

Abstract

Coupling of ultra-high-performance liquid chromatography (UHPLC) with mass spectrometry (MS) offers an efficient alternative to conventional chromatographic separation procedures or simple HPLC-MS. It offers several advantages such as low analysis time, high sensitivity and improved resolution. It's advanced instrumentation and sophisticated working principles have made it the analytical technique of choice in a variety of fields. In the present compilation, a description of the typical instrumentation of UHPLC-MS along with working principles have been provided. Its applications in different areas such as

* Corresponding Author's Email: divya15802@gmail.com.
† Corresponding Author's Email: ad4singh@gmail.com.

In: Modern Approaches in Fluid Chromatography
Editors: Rohit Bhatia and Bhupinder Kumar
ISBN: 979-8-88697-856-8

pharmacokinetics, forensics, drug-metabolites analysis etc. have been elaborated by citing the latest reports including the significant outcomes.

Keywords: liquid chromatography, high-performance, mass spectrometry, sensitivity

1. Introduction

Chromatography is a technique for separating specific components from complicated chemical mixtures [1]. The term "chromatography" (from Greek chroma, color, and graphein, to write) refers to scientific techniques for separating mixtures [2, 3]. It entails passing a mixture dissolved in a "mobile phase" through a stationary phase, which, based on differential partitioning between the mobile and stationary phases, separates the analyte to be measured from other molecules in the mixture [4]. Differences in the partition coefficient of the compounds cause differing retention on the stationary phase and, as a result, separation [5]. One of the most extensively used separation methods nowadays is reversed-phase liquid chromatography (RPLC). Over the last few years, significant advancements in high-throughput studies and efficient separations in RPLC have been made over the previous few years, including unique supports and apparatus. The requirement to handle an escalating number of studies or more complicated samples prompted these advancements [6, 7].

There is an increasing demand for high-throughput separations in several domains, such as toxicology, forensics, clinical studies, and environmental studies. The pharmaceutical industry, with its need for increased production and lower costs, is undoubtedly the primary driver for speedier separations. Due to the high number of analyses required, rapid analytical processes are frequently necessary for popular pharmaceutical applications, such as purity tests, pharmacokinetic investigations, and quality control. Many applications, such as proteomics, plant extract analysis, and metabolomics, require effective separations since they all deal with extremely complex materials, such as biological samples, tryptic digests, or natural plant extracts. Due to the difficulty of these samples, traditional high-pressure techniques are not used.

There are numerous clear limits to liquid chromatography (HPLC) technologies. Ultra-high-pressure liquid chromatography (UHPLC) has quickly established itself as a potent and resilient analytical tool among the several methodologies utilized to accomplish fast and high-resolution separations [8, 9].

In 2005 first UHPLC was commercialized. Afterwards, it was extensively accepted. Other acronyms, such as UPLC (ultra-performance liquid chromatography), VHPLC (very high-pressure liquid chromatography), UFLC (ultra-fast liquid chromatography), or ultra-HPLC, may be used in addition to UHPLC [10]. Ultra-high pressure liquid chromatography (UHPLC) is another term for the process. The van Deemter relationship, which describes the link between flow rate and plate height, is the core concept of UPLC [11]. For good results, the van Deemter equation reveals that the flow range with smaller particles is substantially higher than with bigger particles.

$$H= A+\frac{B}{v}+C\,v\ldots\ldots\ldots \qquad (1)$$

where, H stands for the HETP; A, B & C are the constants and v is the flow rate of the carrier gas. To enhance column efficiency, the goal is to reduce HETP. Eddy mixing is shown by the word A, independent of velocity. It will be smaller if the columns are packed with little, homogenous particles. The word B refers to the particles' inherent inclination to diffuse. This impact is reduced at high flow rates. The word C refers to the kinetic resistance to equilibrium during the separation process. The kinetic resistance is the amount of time it takes to transition from the mobile to the stationary state and again. A molecule on the packing material tends to lag behind molecules in the mobile phase as the flow velocity of the mobile phase increases. As a result, linear velocity and this term are inversely proportional. As a result, the throughput will likely increase, and the separation can be speeded up without compromising chromatographic performance. The column length is proportional to efficiency, whereas the radius of the particles is inversely proportional. As a result, the column length may be lowered by a factor similar to the particle radius without compromising resolution. UPLC has aided in the identification of drug metabolites as well as the improvement of separation spectra quality.

UHPLC has a higher resolution and resolving capacity, making it more selective and sensitive. It also minimizes process cycle time and ensures end-product quality by lowering operating costs and running time [12, 13]. It uses a unique column material with very fine particles to boost sensitivity and speed up analysis. It reduces solvent usage, enhances sample throughput, and allows real-time analysis in conjunction with industrial operations [11, 14]. Apart from so many advantages, there are some demerits, too, compared to standard HPLC; UHPLC has greater back pressures, reducing column life [15]. In UHPLC, increasing the column temperature lowers the back pressure problem. Furthermore, particles smaller than 2 μm are typically non-regenerable and have limited usage [11, 16]. In the presented chapter, the separation powers of UHPLC combined with identification/quantification potentials of mass spectrometry (MS) have been elaborated based on recently reported applications.

2. Instrumentation

The UHPLC systems follow the path of sample injection, UPLC columns and detectors (Figure 1) [17, 18, 19].

The injector accurately measures and adds a tiny amount of solution containing the sample in the mobile phase. The injection must be done consistently and precisely. In UPLC, the sample volume is typically 2-5 μL. For biological material, direct injection techniques are now used. Conventional injection valves can be manual or programmed, and the injection procedure must be somewhat pulse-free to protect the column from excessive pressure instabilities. The swept volume of the device should be as small as possible to avoid potential band spreading. To effectively utilize the speed provided by UPLC, a short injection cycle time is necessary. Low-volume injections with limited carryover are essential to optimize sensitivity.

The solvent delivery system must offer a smooth and consistent flow of solvents at constant pressures. Pumping and injection modules can withstand the high pressure of up to 8000-15000 psi for suitable resolution. The pressure exerted into the column should be pulse free. Aside from the instrument's pressure potential, it's also significant to note that it was modified to function in fast and ultrafast modes with smaller column volumes.

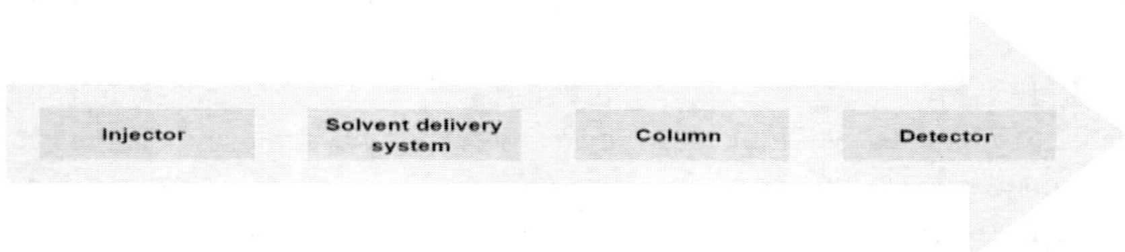

Figure 1. Various parts of the UHPLC system.

Small diameter columns (1 or 2.1 mm I.D.) are widely used in UHPLC. Nonporous particles were utilised exclusively before the development of UHPLC technology. Jorgenson et al. were the first to compare 1.5 mm porous hybrid particles packed into a 30 mm I.D. fused silica capillary column from Waters (Milford, USA) with 1.0 mm nonporous silica material. Hybrid particles performed similarly to 1.0 mm nonporous silica particles in pressure resistance when tested at up to 4500 bar pressures. UPLC columns are filled with particles that are created using different processes such techniques are shown in Figure 2.

CSH Technology is a new approach to hybrid material creation that employs low-level surface-charged particles to improve peak selectivity and sharpness. A hybrid-based packing material strategy delivers sharp peaks with enhanced efficiency and chemical stability, particularly for basic chemicals at low pH. CSH particles are frequently utilized, including CSH C18, CSH Phenyl hexyl, and CSH Fluoro phenyl. These columns feature excellent peak shape, higher loading capacity (CSH C18), selectivity for positional isomers, halogenated, and polar compounds, and selectivity for positional isomers, halogenated, and polar compounds (CSH-fluoro phenyl). Higher stability over a wide pH range increased batch-to-batch repeatability, and quick column equilibration with any change in the mobile phase pH are some of the additional benefits [20].

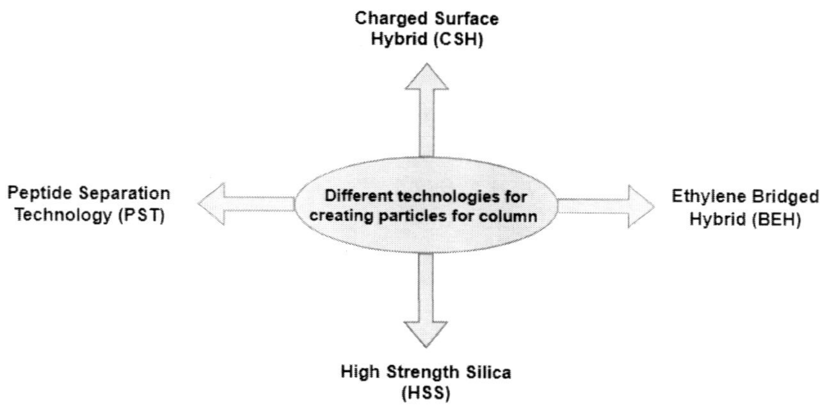

Figure 2. Various types of UHPLC columns.

Another primary driver of UPLC Technology is the Ethylene Bridged Hybrid [BEH] particle. It is suitable for reversed-phase and hydrophilic interaction chromatography and comes in three distinct pore sizes [130, 200, and 300 Å] and numerous bonded phases for a small molecule to big biopharmaceutical characterization. Because of hybrid particle technology's inherent chemical stability, a more extensive pH range [pH 1-12] can be used, making it a versatile and stable separation technology for method development [21].

This results in an enhanced approach that delivers more excellent retention than hybrid particles. HSS T3, HSS C_{18}, HSS C_{18} SB, HSS PFP, and HSS CN are some of the most often utilized HSS particles. These columns provide balanced retention of polar and hydrophobic molecules (T3); excellent peak shapes and increased retention (C18); greater retention of basic compounds (CS_{18} SB: Selectivity for base); ideally suited for planar aromatic, positional, halogenated compounds; ultra-stable retention and compatible with both reversed-phase and normal-phase techniques; ultra-stable retention and compatible with both reversed-phase and normal-phase techniques (PFP) [11].

PST implements C18 BEH Technology particles with particle sizes ranging from 1.7 to 10 m, columns with internal diameters ranging from 75 to 30 mm, and column lengths ranging from 50 to 250 mm. They're used in any peptide-related research and development. Sharp, symmetrical peaks can be found in the PST columns [22].

Detectors- The UPLC detector should be able to provide a high sampling rate with tiny attainable peaks and little peak dispersion so that the separated solute is not wasted on the column. Acquity photodiode array (PDA) and Tunable Vis-UV (TUV) detectors are utilized in the UHPLC, with Teflon AF providing an internally reflecting surface that improves light transmission efficiency by removing internal absorptions. These have 10 nun path lengths, 20 (PDA) and 40 (TUV) acquisition speeds, and a total internal volume of 500 nL. UPLC has also been utilized for mass spectrometric detection. For more than 70 percent of UHPLC separations, UHPLC MS/MS is the detector of choice [12, 23]. This chapter comprises the latest application of UHPLC MS/MS, focusing on the throughput and resolution gains made possible by UHPLC technology.

3. Applications of UHPLC-MS

3.1. Application in Pharmacokinetics

Amikacin is an aminoglycoside antibiotic frequently used to treat severe infections brought on by bacteria that have developed multiple resistances

[37]. Silva et al. showed that Amikacin and Creatinine concentrations in dried plasma spots (DPS) could be determined simultaneously by employing a straightforward and accurate approach that was created and validated using UHPLC-MS/MS. Dried micro samples, such as DPS, were employed to minimize specimen transportation and storage costs in resource-constrained settings, hence expanding access to TDM of AMI. Assays for precision, accuracy, and stability produced satisfactory results. For 14 days at 6°C, 2 days at 22°C, and 1 day at 42°C, AMI remained steady in DPS. At all temperatures that were examined, CRE remained steady for 14 days. By using Passing-Bablok regression and Bland and Altmann graphs to explore the AMI and CRE concentrations in DPS and plasma, equivalent findings were obtained. Using DPS or plasma concentrations, estimates of the patient's clearance, distribution volume, and indicated AMI dosages were comparable. The test offers a practical logistic alternative that will enable more people to have access to dosage individualization of AMI in environments with constrained resources [24] (Table 1).

JIB-04 is a small drug that prevents HIV-1 Tat expression, transactivation, and viral replication by pan-selectively inhibiting SHMT2 in T cell lines. It also inhibits the demethylase activity of KDMs *in vitro*, in cancer cells, and *in vivo* [25]. Wang et al. developed and validated an LC-MS/MS technique for measuring JIB-04 in rat plasma samples that is sensitive, specific, quick, and reliable. Researchers used a SCIEX6500+ triple QUAD LC-MS system outfitted with an ExionLC UHPLC unit to design and thoroughly validate the technique. The chromatographic separation was accomplished on a reverse phase ACE Excel 2 Super C18 column with a flow rate of 0.5 mL/min under gradient elution. Over concentration ranges of 0.5 to 1000 ng/mL, the calibration curves were linear ($r^2 > 0.999$). The precision (CV percent) was 10.2 percent or less, while the accuracy (RE percent) ranged from -7.4 to 3.7 percent (Table 1). The stability results demonstrated that no appreciable deterioration took place during the experiment. The oral bioavailability of JIB-04 was determined to be 44.4 percent when this approach was effectively used in the pharmacokinetic analysis of JIB-04 in rat plasma following intravenous and oral administration.

Due to its numerous medical benefits, aloe plants have been used as folk remedies for a very long time. Due to its calming qualities, aloe has been widely used in cosmetic products to aid in healing skin injuries. Additionally, it has been demonstrated that aloe has anti-infection, anti-inflammatory, and laxative benefits people. Anthraquinone-C-glycoside Aloin-A (also known as barbaloin), which is generated from Aloe latex, is one of the primary bioactive ingredients and one of the chemotaxonomic indicators of many Aloe species, including Aloe vera [38] (Table 1).

As it lowers iNOS expression and NF-B activity, Aloin-A is said to have anti-oxidative and anti-inflammatory properties. It also has antineoplastic properties [39]. Aloin-A also shows immunomodulatory, hepatoprotective, and antiviral properties [40].

The current work devised a quick technique for quantifying aloin-A in rat plasma using ultra-high performance liquid chromatography-tandem mass spectrometry (UHPLC-MS/MS). Using obtusin as an internal standard (IS), the plasma was prepared in a single step, and the analytes were separated using an Agilent C18 column with a mobile gradient phase made of acetonitrile and formic acid aqueous solution. Aloin-A and IS's multiple reaction monitoring transitions were m/z 417.1 and 297.0 for negative ion electrospray and m/z 343.1 and 328.1 for IS, respectively. Therefore, the devised technique was verified with a linear range of 1–1000 ng/mL (Table 1). According to FDA instructions, every validation parameter was within the acceptable range. After intravenous and oral treatment, samples from a pharmacokinetic investigation in healthy rats were successfully analyzed using the established methodology. It was discovered that aloin-A was immediately absorbed, widely disseminated, and swiftly removed. Aloin-A's absolute bioavailability was found to be 5.79 percent [26].

The dried ripe seed of *Cassia tora L.* and *Cassia obtusifolia L.* is known as semen Cassiae in Chinese [41]. In China, it has been used professionally for millennia to relieve constipation, enhance vision, positively impact the liver, and avoid hyperlipidemia [42]. It can be taken alone or as part of a composite prescription. Science has progressively discovered Semen Cassiae's anti-inflammatory, antioxidant, and attenuating diabetic nephropathy properties [43]. Numerous anthraquinone chemicals have already been isolated from Semen Cassiae and demonstrated to be bioactive by researchers, including emodin, rhein, chrysophanol, physcion, aloe-emodin, obtusi-folin, chryso-obtusin, obtusin, and aurantio-obtusin because Semen Cassiae has a wide range of medicinal applications and is safe to use.

Semen Cassiae is widely used in various goods, including traditional, tea, health products, and ultrafine granular powder (UGP). Li Xie et al. developed a quick, sensitive, and trustworthy UHPLC-MS/MS technique and validated it by the FDA's most recent bioanalytical method recommendations.

Table 1. Reported applications of UHPLC-MS in pharmacokinetics

S. No	Analytes	Matrix	Condition	Run time	R^2 value	LOD	LLOQ	Ref
1	Amikacin	Dried plasma spot	Acquity®HSS T3 (100 × 2.1 mm, p. d. 1.8 μm) column at 40°, flow rate 0.4 mL	5.5 min	0.994		5.1	[24]
2	JIB-04	Rat Plasma	ACE Excel 2 Super C18, flow rate 0.5 mL/min	3 min	0.999	0.05 ng/mL	0.5 ng/mL	[25]
3	Aloin-A	rat plasma	Agilent ZORBAX Eclipse Plus C18 column (2.1 × 50 mm, 1.8 μm)	2.5 min	0.990	0.5 ng/mL	1 ng/mL	[26]
4	Nine anthraquinones of Semen Cassiae	Rat	AgilentExtend-C18column (100 mm × 2.1 mm, 1.8 μm), 0.3 mL/min, 40°C	8 min	0.9989 0.9992 0.9934 0.9981 0.9969 0.9939 0.9952 0.9956 0.9977		0.313 ng/mL 0.320 ng/mL 1.000 ng/mL 0.318 ng/mL 0.313 ng/mL 0.783 ng/mL 0.395 ng/mL 0.802 ng/mL 0.400 ng/mL	[27]
5	*Hedyotis diffusa* Willd extract	Rat plasma	Agilent ZORBAXRRHD Eclipse Plus C18 column (2.1 mm × 50 mm, 1.8 μm), flow rate 0.3 mL/min.		> 0.998		30 ng/mL 15 ng/mL 80 ng/mL 30 ng/mL 10 ng/mL 30 ng/mL 2.5 ng/mL 1.5 ng/mL 15 ng/mL 75 ng/mL	[28]

Table 1. (Continued)

S. No	Analytes	Matrix	Condition	Run time	R^2 value	LOD	LLOQ	Ref
6	Venetoclax	mouse plasma	Accucore aQ column (50 mm × 2.1 mm, dp =2.6 μm, Thermo Fisher Scientific) with a C18 AQUASIL guard cartridge (2.1 mm × 10 mm, dp =3 μm, Thermo Fisher Scientific)	4.6 min	0.995		5 ng/mL	[29]
7	Cefazolin Cefalothin	Human plasma	Kinetex C8, 2.1 × 50 mm, 1.7 μm, 100 Å column from Phenomenex, 50°C	5min	0.9999 0.9991	1 μg/mL	1 μg/mL	[30]
8	Apixaban, Dabigatran Edoxaban, Rivaroxaban	Human Serum	Acquity UPLC HSS T3 1.8 μm, 2.1 × 100 mm analytical column and an Acquity UPLC HSS T3 1.8 μm, 2.1 × 5 mm pre-column, 50°C	3.0 min	0.996 0.999 0.996 0.998	0.25ng/mL 0.5 ng/mL 2.0 ng/mL 0.5 ng/mL	2.5 ng/mL 2.5 ng/mL 5.0 ng/mL 2.5 ng/mL	[31]
9	Ziyuglycoside I	Leucopenia rat	Shim-pack GIST column (2.1 mm I.D.×50 mm l, 2.0 μm), 40°C	5 min	0.999			[32]
10	Cortex Dictamni extract	Rat plasma	Thermo Syncronis C18 column (2.1 mm × 100 mm, 1.7 μm)	9 min	0.9990 0.9981 0.9981 0.9980		9.18 ng/mL 12.00 ng/mL 16.05 ng/mL 4.59 ng/mL	[33]

S. No	Analytes	Matrix	Condition	Run time	R^2 value	LOD	LLOQ	Ref
11	Six flavonoids, gallic acid and 5,8-di hydroxy-1,4-naphtho-quinone	Rat plasma	VenusilASB C18 column (2.1 mm × 100 mm, 5.0 μm), 20°C	14 min	0.9927 0.9918 0.9940 0.9977 0.9979 0.9959 0.9984 0.9983		20 ng/mL 5 ng/mL 3 ng/mL 10 ng/mL 6 ng/mL 3 ng/mL 2 ng/mL 100 ng/mL	[34]
12	Six bioactive constituents of Guizhi Fuling Capsule	Rat Plasma	Waters ACQUITY UPLC BEHC18column (50 mm × 2.1 mm, 1.7μm), 0.2 mL/min, 30°C	6.0 min	0.9996 0.9980 0.9955 0.9973 0.9972 0.9971		10 ng/mL 10 ng/mL 5 ng/mL 5 ng/mL 25 ng/mL 25 ng/mL	[35]
13	Cinnam-aldehyde, cinnamic acid, and 2-methoxy cinnamic acid	Rat whole blood	Thermo Scientific C18 column (2.1 mm × 50 mm, 1.9 μm), 0.2 mL/min, RT	4 min	0.9941 0.9954 0.9965		0.1 ng/mL 5.8 ng/mL 10.0 ng/mL	[36]

The comparative pharmacokinetics of nine active components of Semen Cassiae in rat plasma were assessed using the established technique. The quantification of the analytes was carried out using electrospray ionization in multiple reaction monitoring modes. The separation was performed using ultra-pure water and acetonitrile as the mobile phase on an Agilent Extend-C18 column (100 mm 2.1 mm, 1.8 m) (Table 1). Ethyl acetate was used for liquid-liquid extraction to prepare the samples [27].

Traditional Chinese medicine *Hedyotis diffusa* Willd (*H. diffusa)* has long been utilised in folk remedies and is extensively available in east and Southeast Asia, including China, Japan, and Indonesia [44]. It includes trace elements and amino acids in addition to anthraquinones, terpenes, flavonoids, organic acids, alkaloids, sterols, alkanes, polysaccharides, *Hedyotis diffusa*, and other components. These ingredients have been found to have therapeutic properties such as anti-cancer, anti-fibroblast, immunomodulatory, and neuroprotective effects [44]. For instance, they can be used to treat hepatitis, tonsillitis, sore throat, appendicitis, urethral infections, and malignant liver, lung, and stomach tumors as well as anti-oxidant and anti-inflammatory conditions. For the simultaneous determination of fourteen bioactive ingredients (gallic acid, geniposidic acid, protocatechuic acid, caffeic acid, ferulic acid, scopoletin, apigenin-7-o-glucuronide, daidzein, apigenin, ursolic acid, oleanolic acid, sitosterol, coniferin, and stigmasterol) in the plasma, Chen et al. developed and validated. Internal ultra-high benchmarks included Danshensu and Icarin (IS1 and IS2). Utilizing an Agilent ZORBAXRRHD Eclipse Plus C18 column (2.1 mm 50 mm, 1.8 m) and gradient elution with mobile phase made up of 0.1 percent acetic acid water (solvent A) and methanol (solvent B) and pumped at a flow rate of 0.3 mL/min, the chromatographic separation was carried out (Table 1). Electrospray ionization (ESI) in positive and negative modes was used for mass spectrometric detection in the multiple reaction monitoring (MRM) modes. Organ homogenates were processed by solid-phase extraction (SPE) with Waters Oasis HLB 3 cc (60 mg), whereas plasma samples were prepared by protein precipitation with 300 L of methanol containing 0.1 percent (v/v) formic acid [28].

An oral BH3-mimetic called venetoclax (ABT-199, Venclexta) inhibits the anti-apoptotic B-cell lymphoma-2 (BCL-2) protein. A component of both first-line and relapsed/refractory therapy regimens for chronic lymphocytic leukaemia and acute myeloid leukaemia, venetoclax has revolutionised the treatment of hematologic malignancies [45]. Eric D et al. Thermo Fisher Scientific's Quantiva triple quadrupole mass spectrometer and Vanquish

UHPLC were utilised for the analysis. For analyte separation, a C18 AQUASIL guard cartridge (2.1 mm 10 mm, dp = 3 m, Thermo Fisher Scientific) was used with an Accucore aQ column (50 mm 2.1 mm, dp = 2.6 m). The autosampler was kept at 4°C and the column at 20°C. Gradient elution was carried out for 4.6 min at a flow rate of 0.4 mL/min using solvent A (0.1 percent FA in H_2O) and solvent B (0.1 percent FA in ACN-MeOH, 50:50 v:v) as the mobile phase. The gradient conditions were as follows: 0 to 1.99 min, 10% B; 2.0 to 3.5 min, 65 to 99% B; 3.5 to 4.0 min, 95% B; 4.1–4.6 min, 10% B. In this research, it was reported that the developed and validated method was a simple, rapid, and accurate UHPLC-MS/MS method for determining venetoclax in mouse plasma [29] (Table 1).

First-generation cephalosporin-lactam medicines like cefazolin and cephalothin are widely used to treat sepsis and other life-threatening infections [46]. Peritoneal dialysis, a common renal replacement therapy for those with severe chronic kidney disease, frequently leads to peritonitis, which accounts for more than 18% of infection-related mortality in peritoneal dialysis patients [47]. An analytical UHPLC-MS/MS technique was created by Parker et al. to assess cephalothin or cefazolin in patient plasma (total and unbound), urine, and peritoneal dialysate. The determination of unbound levels of cephalothin in plasma and peritoneal dialysate was described initially using this approach. This approach has the benefits of easy sample preparations, small sample sizes, great precision and accuracy, repeatable recoveries, and little matrix effects. The pharmacokinetic profiles of individuals with advanced chronic kidney disease who were taking cephalothin or cefazolin while receiving renal replacement therapy may be characterized using this technique. Cefazolin and cephalothin in plasma (total and unbound), urine, and peritoneal dialysate were separated using gradient elution with mobile phases made up of (a) 0.1 percent formic acid and (b) 0.1 percent formic acid in methanol on a Kinetex C8, 2.1 50mm, 1.7 m, 100 columns from Phenomenex (operated at 50°C). A 2.1mm UHPLC C18 Security Guard ULTRA protection cartridge was used to protect the column (Phenomenex). A methanol concentration gradient was used to control the mobile phase's operation. At the beginning of the run, the column was equilibrated with 5 percent methanol in 0.1 percent formic acid. From 1.5 to 2 minutes later, the methanol concentration increased to 35 percent, and from 2.5 to 3.5 minutes later, it increased once more to 55 percent. Then, at 4.1 minutes, the methanol content was back to 5 percent till the end of the run of 5 min. For LC-MS/MS analysis, a positive-ion electrospray ionization source interface was employed. The interface parameters were 3

L/min of nebulizing gas flow, 250°C for the desolvation line, 400°C for the heat block, and 15 L/min for the drying gas flow [30] (Table 1).

The gold standard for prophylaxis and treatment of venous thromboembolic illness has historically been anticoagulation therapy with the vitamin K antagonist warfarin in individuals with atrial fibrillation [48]. However, anticoagulants with a more direct mode of action have been created within the past ten years. Among these medications, dabigatran, a direct thrombin inhibitor, and the factor Xa inhibitors rivaroxaban, apixaban, and edoxaban. These medications are frequently referred to as "direct oral anticoagulants" (DOACs), "new oral anticoagulants," or "non-vitamin K antagonist oral anticoagulants" as a collective term (NOACs) [49]. Lindahl et al. used Acquity UPLC HSS T3 1.8 m, 2.1 100 mm analytical column system for separation. 1 μL injection volume was chosen. The mobile phase included acetonitrile (B) flowing at 0.6 mL/min and MilliQ H_2O with 0.1 percent formic acid (A). In 3.0 minutes, the samples were eluted using a linear gradient from 95% A to 98% B. The autosampler temperature was 10°C, whereas the column temperature was set at 50°C. Methanol, acetonitrile, 2-propanol, MilliQ H_2O, and formic acid (25:25:25:23:2, v/v) were used for pre-and post-inject washes, lasting 2 and 6 seconds, respectively. It was carried out using a Z-spray electrospray interface-equipped tandem-quadrupole mass spectrometer, the Xevo TQ-S. The multiple reaction monitoring (MRM) methods of positive electrospray ionization were used. The source block temperature was set to 120°C, the desolvation gas (nitrogen) was heated to 650°C and given at a flow rate of 1000 L/hour, and the cone gas (nitrogen) was set to 150 L/hour. The capillary voltage was set at 1.0 kV. When there were 15 to 20 data points per peak, dwell durations were automatically modified [31] (Table 1).

Sanguisorba officinalis L., a traditional Chinese herbal remedy, has been used therapeutically for more than ten years to treat radiotherapy- and chemotherapy-induced leucopenia [50]. It was once used extensively for cooling blood and hemostasis. The primary active ingredient of *Sanguisorba officinalis L.* is ziyuglycoside I (ZgI), a pentacyclic triterpenoid saponin [51]. Since 2008, when ZgI's pharmacological action as an antiwrinkle agent was initially discovered, interest in this activity has increased. Later research has demonstrated that ZgI, like *Sanguisorba officinalis L.*, improves bone marrow depression in leukopenic rats and raises white blood cells (WBCs) levels and platelets. To detect ZgI, Zhu et al. have devised a quick and accurate UHPLC-MS/MS approach. On applying this technique to a fresh model of cyclophosphamide-induced leukopenia. Shimadzu Chromatographic System with LC-30AD 2 infusion pumps, DGU-20A5 online degasser, SIL-30AC automated sampling device, CTO-20A column oven, CBM-20A system controller, and an LC-MS-8050 triple-quadrupole mass spectrometer was used to conduct the UHPLC-MS/MS analysis. The devised approach was a quick and accurate detection technique utilized to examine the ZgI pharmacokinetic properties in leukopenic and normal rats. Their findings suggested that the pharmacokinetic properties of ZgI, with shorter T_{max}, decreased C_{max}, and extended t1/2, might be changed in the leukopenic state. These findings starkly contrast with past studies on ZgI's pharmacokinetics in a disease-free state [32] (Table 1).

Many Asian and European nations extensively employ Cortex Dictamni, the root cortex of Dictanmnus dasycarpus Turcz (Rutaceae), as a traditional remedy. It is frequently advised to treat various conditions, including eczema, jaundice, rubella, tinea, scabies, acute rheumatism arthritis, and coronary atherosclerosis. More than 100 chemical components, including limonoids, furoquinoline alkaloids, flavonoids, coumarins, sesquiterpene, and sesquiterpene glycosides, have been identified from Cortex Dictamni after extensive scientific research. Four critical components of the Cortex Dictamni extract—limonin, dictamnine, obacunone, and fraxinellone—are examples of the principal bioactive chemicals among them: limonoids and furoquinoline alkaloids. The four constituents' documented pharmacological actions are listed below. Limonin has been demonstrated to have significant neuroprotective effects against glutamate-induced neurotoxicity in primary cultures of rat cortical cells. Dictamnine has also shown significant effects on anti-platelet aggregation, vascular relaxation, and suppressing Ca^{2+} influx through both voltage-dependent and receptor-operated Ca^{2+} channels. Obacunone was also discovered to significantly increase the cytotoxicity of microtubule inhibitors like vinblastine and taxol. According to reports, fraxinellone has cytotoxic, antimicrobial, antitumor-promoting, anti-inflammatory, and antibacterial properties. Wang et al. devised an extremely high-performance liquid chromatography-tandem mass spectrometry approach for the simultaneous measurement of four key constituents. Analytes were separated on a Thermo Syncronis C18 column by a gradient elution within a runtime of 9 min. A (methanol) and B (2 mmol/L ammonium acetate in water) made up the mobile phase. Positive ion electrospray ionization in multiple reaction monitoring modes was used to perform the detection. Following oral administration of Cortex Dictamni extract, the pharmacokinetics of limonin, dictamnine, obacunone, and fraxinellone in rat plasma were effectively studied using this quick and sensitive approach [33] (Table 1).

Traditional Chinese medicine has long employed Cortex *Juglandis mandshuricae* as a folk cure to treat cancer, diarrhoea, and dysentery [52]. Major bioactive components of Cortex *Juglandis mandshuricae* extract include gallic acid and 5,8-dihydroxy-1,4-naphthoquinone, as well as six flavonoids (myricitrin, quercitrin, taxifolin, myricetin, quercetin, and naringenin) [53]. Using chloromycetin as an internal standard, ultrahigh-performance liquid chromatography and the tandem mass spectrometry approach were developed to simultaneously determine eight components in rat plasma. Hydrochloric acid was used to acidify plasma samples with vitamin C (antioxygen) added before ethyl acetate was used for liquid-liquid extraction. Eight chemicals were separated on a Venusil ASB C18 column and then identified using the multiple reaction monitoring modes and electrospray ionization in the negative ion mode. The technique was linear for all analytes within the examined range, and all correlation coefficients were higher than 0.9900. Gallic acid's confirmed lower limit of quantification was 20 ng/mL, myricitrin's was 5 ng/mL, quercetin was 3 ng/mL, taxifolin's was 10 ng/mL, myricetin's was 6 ng/mL, naringenin's was 2 ng/mL, and 5,8-dihydroxy-1,4-naphthoquinone' Accuracy (RE percent) varied from 6.9 percent to 6.9 percent, and intra- and inter-day precisions (RSD percent) were less than 15 percent. Rats were given an oral dose of Cortex *Juglandis mandshuricae* extract, and the pharmacokinetics of the eight analytes were investigated using the validated technique Sun et al. [34] (Table 1).

Zhao et al. in their exploratory work used Guizhi Fuling Capsule. A significant and popular Chinese patent medication used to treat gynaecological blood stagnation, including endometriosis, hysteromyoma, dysmenorrhea, and oophoritic cyst. For the simultaneous determination of gallic acid, amygdalin, albiflorin, paeoniflorin, paeonol, and cinnamic acid, the main bioactive components of Guizhi Fuling Capsule in rat plasma using phenacetin as internal standard, an ultra-high performance liquid chromatography-tandem mass spectrometry (UHPLC-MS/MS) method was created (IS). After pretreatment with acetonitrile-based protein precipitation and acidification, the plasma samples were separated on a Waters BEH C18 column (50 mm 2.1 mm, 1.7 m) using gradient elution with a mobile phase made up of water (containing 0.1 percent formic acid) and acetonitrile at a flow rate of 0.2 mL/min. On a Micromass Quattromicro API mass spectrometer outfitted with an electrospray ionization source, mass spectrometric detection was carried out in multiple reaction monitoring (MRM) modes. For all analytes, the intra- and inter-day precisions (measured as relative standard deviation) were less than 14.6%, and the accuracies (measured as relative error) were within 5.0%.

The lowest limits of quantification (LLOQ) for gallic acid, amygdalin, albiflorin, paeoniflorin, paeonol, and cinnamic acid were 10, 10, 5, 5, 25, and 25 ng/mL, respectively. Rat plasma demonstrated good extraction recovery, matrix effect, and stability. Following oral administration of Guizhi Fuling Capsule to rats, this approach was thoroughly verified and used to conduct a pharmacokinetic evaluation of the six bioactive ingredients [35] (Table 1).

According to the Chinese Pharmacopoeia, Cinnamomi Ramulus, the stem of *Cinnamomum cassia* Presl, is one of the most significant Chinese medicinal substances. Due to its special qualities, including anti-inflammation, antiviral, antibacterial, and antioxidant activity [54], Cinnamoni Ramulus essential oil (CRO) has been widely utilized as a pharmaceutical, food, and cosmetic ingredient. Cinnamaldehyde, cinnamic acid, 2-methoxy cinnamic acid, coumarin, cinnamic alcohol, and 2-methoxy cinnamaldehyde are only a few aromatic substances found in significant levels in CRO. The main bioactive chemicals among them were identified as cinnamaldehyde, cinnamic acid, and 2-methoxy cinnamic acid. The most prevalent component, cinnamonaldehyde, has been shown to have anti-stress urinary incontinence, anti-inflammation, anti-thrombotic, and anti-tumor properties. It had been thought that cinnamic acid possessed antibacterial, antioxidative, and anticancer properties. It was asserted that 2-methyl cinnamic acid has strong antioxidant and antityrosinase properties. Cinamaldehyde, cinnamic acid, and 2-methoxy cinnamic acid were all simultaneously determined in rat whole blood using a straightforward and quick ultra-high performance liquid chromatography-tandem mass spectrometry (UHPLC-MS/MS) approach by Ji et al. it was the first time 2-methoxy cinnamic acid's pharmacokinetics in rat whole blood had been studied. Samples underwent a one-step protein precipitation procedure using acetonitrile and 37 percent formaldehyde (90:10, v:v). Chromatographic separation was carried out at ambient temperature on a Thermo Scientific C18 column (2.1 mm 50 mm, 1.9 m). It lasted for four minutes in total. The detection was carried out using multiple reactions monitoring mode electrospray ionization with positive and negative ions [36] (Table 1).

3.2. Application of UHPLC-MS/MS in Forensics

New psychoactive substances (NPSs) have recently appeared on the black market. NPSs continue to be a concern to public health even though several nations are working to reverse this trend. For example, more than 170 NPSs are presently under control in China. Most NPSs are designer cathinone and synthetic cannabinoids. However, there has been a significant rise in the use of innovative synthetic opioids, primarily fentanyl and its variants [62]. Fentanyl, an agonist of the -opioid receptor, is 50–100 times more potent than morphine, cocaine, and methamphetamine when used as an analgesic and anaesthetic adjuvant [63].

Due to an increase in mortality, and the introduction of newer fentanyl analogs, and synthetic opioids, to identify 37 fentanyl analogs and new synthetic opioids in hair samples, Qin et al. devised an analytical technique and as well validated it using UHPLC-MS/MS. Cryogenic grinding was used to extract hair samples (20 mg) from an extraction solution containing methanol, acetonitrile, and 2 mmol/L ammonium acetate (pH 5.3). The analytes were separated using a WATERS Acquity UPLC HSS T3 column connected to a QTRAP 6500 PLUS triple quadruple linear ion trap mass spectrometer after the samples were centrifuged. Mobile phase A, a 20 mmol/L ammonium acetate solution containing 0.1 percent formic acid and acetonitrile, made up the mobile phase (mobile phase B). The autosampler's temperature was set to 4°C, and the injection volume was 5L. Additionally, the technique was effectively used to identify and measure fentanyl and sufentanil in hair samples from two real instances. Therefore, this technique has much potential for finding new synthetic opioids and fentanyl analogs in forensic investigations [55] (Table 2).

Hair plays a crucial role in drug-related crimes as a substitute for or in addition to urine and blood (DFC). Hair concentrations can be low, even below the pg/mg range, and single dosages are frequently given. Wang et al. employed tandem mass spectrometry to identify drugs at these low levels. Using a multi-target technique, a wide range of substances are covered in a single examination, guaranteeing an efficient approach to forensic toxicology. A Xevo TQS (Waters) tandem mass spectrometer was used for detection. Both positive and negative ionization modes were used. For 51 percent of the essential drugs, the lower limits of quantification (LLOQs) were in the range of 0.05–0.5 pg/mg, and the majority (98 percent) were ≤ 5 pg/mg. Linearity ranged from LLOQs to 100–500 pg/mg for all basic drugs. The LLOQs ranged from 0.4 to 500 pg/mg for acid and neutral drugs, and linearity ranged from LLOQs to 80–40,000 pg/mg. The developed method was applied to forensic cases and concentrations of rarely reported drugs in hair in 25 postmortem forensic cases were presented. Hair concentrations of amisulpride, gabapentin, mianserin, mepyramine, orphenadrine, and xylometazoline have not been previously reported [56] (Table 2).

The most time-consuming stage in bioanalysis is sample preparation, which often involves many procedures and accounts for around 80% of the total analysis time. Blood samples have traditionally been prepared using liquid-liquid extraction, protein precipitation, solid-phase extraction, and solid-phase microextraction. Guthrie and Susi employed the dried blood spot technique (DBS) on human blood for the first time to diagnose phenylketonuria in sizable groups of newborn newborns in 1963 [64]. DBSs were utilized in epidemiological, toxicokinetics, and pharmacokinetic investigations, diagnostic screening, and therapeutic medication monitoring, in addition to neonatal screening for metabolic abnormalities. To simultaneously identify and quantify illegal substances and their metabolites utilizing DBS on postmortem blood samples, Odoardi et al. set out to evaluate a UHPLC-tandem mass spectrometry approach. Blood samples were blended with internal standards before being spotted on the card in this exploratory study to reduce spotting variability that might impact quantitative results. The technique was tested using actual blood samples from autopsy cases. A binary pump with an integrated vacuum degasser, a high-performance well-plate autosampler, and thermostated column compartment modules made up the Agilent 1290 Infinity UHPLC system. A triple quadrupole mass spectrometer with a jet stream electrospray ionisation source was used as the detection equipment. A C18 column that was only slightly permeable was employed for this investigation.

Only 100 μL of the sample, along with internal standards, was needed for DBS extraction. Three droplets (30 μL each) of this solution were then spotted on the card, dried for an hour, punched, and extracted with methanol containing 0.1 percent formic acid. After the supernatant was evaporated, the residue was then reconstituted in 100 μL of water with 0.1 percent formic acid and added to the UHPLC-MS/MS apparatus. The devised technique was verified by considering the factors: matrix effect, dilution integrity, linearity, precision, and accuracy. LOQs were 0.2-2 ng/mL and LODs ranged from 0.05 to 1 ng/mL. The approach demonstrated good linearity for all compounds, with determination coefficients that were consistently greater than 0.99. The tested procedure was then used on an actual postmortem blood sample [57] (Table 2).

Table 2. Reported applications of UHPLC-MS/MS in Forensic Sciences

S. No	Analytes	Matrix	Condition	Run time	R2 value	LOD	LLOQ	Ref.
1	37 fentanyl	Hair	Acuity UPLC system		0.99672	0.5pg/mg	2pg/mg	[55]
	analogues	Specimens	HSS T3 Column		0.99461	1 pg/mg	5 pg/mg	
			(100 mm × 2.1 mm,		0.99417	05 pg/mg	2 pg/mg	
			1.8 μm) fitted with a		0.99541	0.5 pg/mg	2 pg/mg	
			1.8 μm HSS T3		0.99204	2.5 pg/mg	5 pg/mg	
			guard column,		0.99664	0.5 pg/mg	2 pg/mg	
			flow rate 0.2 mL/min,		0.99698	0.5 pg/mg	2 pg/mg	
			4°C		0.99481	0.5 pg/mg	2 pg/mg	
					0.99566	2.5 pg/mg	5 pg/mg	
					0.99606	2.5 pg/mg	5 pg/mg	
					0.99688	2.5 pg/mg	5 pg/mg	
					0.99069	2.5 pg/mg	5 pg/mg	
					0.99740	1 pg/mg	5 pg/mg	
					0.99788	1 pg/mg	5 pg/mg	
					0.99776	1 pg/mg	5 pg/mg	
					0.99636	1 pg/mg	5 pg/mg	
					0.99710	2.5 pg/mg	5 pg/mg	
					0.99744	2.5 pg/mg	5 pg/mg	
					0.99756	1 pg/mg	5 pg/mg	
					0.99676	2.5 pg/mg	5 pg/mg	
					0.99379	1 pg/mg	5 pg/mg	
					0.99379	0.5 pg/mg	2 pg/mg	
					0.99345	0.5 pg/mg	2 pg/mg	
					0.99618	0.5 pg/mg	2 pg/mg	
					0.99564	0.5 pg/mg	2 pg/mg	
					0.99606	2.5 pg/mg	5 pg/mg	
					0.99648	0.5 pg/mg	2 pg/mg	
					0.99499	0.5 pg/mg	2 pg/mg	

S. No	Analytes	Matrix	Condition	Run time	R2 value	LOD	LLOQ	Ref.
1 (cont'd)					0.99686 0.99222 0.99537 0.99632 0.99714 0.99570 0.99744 0.99505 0.99606	2.5 pg/mg 0..5 pg/mg 1 pg/mg 0.5 pg/mg 0.5 pg/mg 1 pg/mg 0.5 pg/mg 0.5 pg/mg 2.5 pg/mg	5 pg/mg 2 pg/mg 5 pg/mg 2 pg/mg 2 pg/mg 5 pg/mg 2 pg/mg 2 pg/mg 5 pg/mg	
2	Broad range of drugs of abuse	human hair	ACQUITY UPLC HSS C18 column (150 × 2.1 mm, 1.8 μm), 0.4 mL/min, 50°C	19.75 min			0.05–0.5 pg/mg for basic drugs, 0.4 to 500 pg/mg for acidic drugs	[56]
3	Dried blood spot	cadaveric blood	Kinetex C18 column (2.6 μm, 100 × 2.1 mm), 300 μL/min, 40°C,		0.99	0.5 ng/mL 0.05 ng/mL 0.5 ng/mL 0.5 ng/mL 0.5 ng/mL 0.5 ng/mL 1 ng/mL 0.05 ng/mL 0.5 ng/mL 0.5 ng/mL 0.5 ng/mL 0.5 ng/mL 0.5 ng/mL 0.5 ng/mL 0.5 ng/mL 0.05 ng/mL	2 ng/mL 0.2 ng/mL 2 ng/mL 2 ng/mL 2 ng/mL 2 ng/mL 2 ng/mL 0.2 ng/mL 2 ng/mL 2 ng/mL 2 ng/mL 2 ng/mL 2 ng/mL 2 ng/mL 2 ng/mL 0.2 ng/mL	[57]

Table 2. (Continued)

S. No	Analytes	Matrix	Condition	Run time	R2 value	LOD	LLOQ	Ref.
3 (cont'd)						0.5 ng/mL 0.5 ng/mL 0.05 ng/mL	2 ng/mL 2 ng/mL 0.2 ng/mL	
4	28 fentanyl analogue	whole blood	Acquity TM UPLC HSS T3 column (100 mm × 2.1 mm, 1.8 μm), 0.3 mL/min, 4°C	8 min.	0.99277 0.99353 0.99299 0.99509 0.99489 0.99003 0.99269 0.99056 0.99190 0.99664 0.99144 0.99337 0.99247 0.99014 0.99315 0.99202 0.99231 0.99790 0.99331 0.99251 0.99481 0.99620 0.99495 0.99622	0.03ng/ml 0.02 ng/ml 0.03 ng/ml 0.03 ng/ml 0.03 ng/ml 0.02 ng/ml 0.03 ng/ml 0.02 ng/ml 0.03 ng/ml 0.02 ng/ml 0.03 ng/ml 0.02 ng/ml 0.03 ng/ml 0.02 ng/ml 0.02 ng/ml 0.02 ng/ml 0.02 ng/ml 0.02 ng/ml 0.03 ng/ml 0.03 ng/ml 0.03 ng/ml 0.1 ng/ml 0.01 ng/ml 0.01 ng/ml	0.2ng/ml 0.05 ng/ml 0.05 ng/ml 0.2 ng/ml 0.05 ng/ml 0.05 ng/ml 0.05 ng/ml 0.05 ng/ml 0.05 ng/ml 0.05 ng/ml 0.05 ng/ml 0.05 ng/ml 0.05 ng/ml 0.05 ng/ml 0.05 ng/ml 0.05 ng/ml 0.05 ng/ml 0.05 ng/ml 0.05 ng/ml 0.2 ng/ml 0.2 ng/ml 0.05 ng/ml 0.05 ng/ml 0.05 ng/ml	Nan et al.

S. No	Analytes	Matrix	Condition	Run time	R2 value	LOD	LLOQ	Ref.
4 (cont'd)					0.99495 0.99361 0.99335 0.99782 0.99782	0.01 ng/ml 0.01 ng/ml 0.01 ng/ml 0.005 ng/ml 0.01 ng/ml	0.05 ng/ml 0.05 ng/ml 0.05 ng/ml 0.05 ng/ml 0.05 ng/ml	
5	16 tryptamines and their metabolites	Human hair	Acquity UPLC HSS T3 column (100 mm × 2.1 mm, 1.8 μm, 0.3 mL/min, 4°C	11 min	0.992 0.995 0.992 0.995 0.992 0.992 0.993 0.995 0.992 0.995 0.994 0.992 0.993 0.993 0.998 0.992	3pg/mg 3 pg/mg 3 pg/mg 3 pg/mg 50 pg/mg 3 pg/mg 10 pg/mg 3 pg/mg 50 pg/mg 3 pg/mg 10 pg/mg 10 pg/mg 10 pg/mg 3 pg/mg 10 pg/mg 10 pg/mg	0.1 pg/mg 0.1 pg/mg 0.1 pg/mg 1 pg/mg 5 pg/mg 1 pg/mg 5 pg/mg 0.1 pg/mg 20 pg/mg 0.3 pg/mg 5 pg/mg 5 pg/mg 1 pg/mg 0.1 pg/mg 0.3 pg/mg 5 pg/mg	[58]
6	GHB, precursors and metabolites	plasma, urine, cerebrospinal fluid and hair	C18 column (2.1 mm × 150 mm, 1.8 μm), 0.3 mL min^{-1}		0.9991-0.9999	0.1-0.3 μg mL^{-1}	0.4-1.0 μg mL^{-1}	[59]

Table 2. (Continued)

S. No	Analytes	Matrix	Condition	Run time	R2 value	LOD	LLOQ	Ref.
7	Cannabinoids	Oral Fluids	Kinetex C18 column (2.6 µm, 100 × 2.1 mm), 400 µL/min	8 min.	0.999 0.996 0.998	2 ng/mL 20 ng/mL	2 ng/mL 10 ng/mL	[60]
8	87 NPS and 32 other drugs of abuse	Hairs and nails	Oasis HLB (5 µm 4.6 × 20 mm), 1 mL/min, 50°C.	10 min	≥ 0.99		20000 pg compound for milligram hair 1000 pg compound per milligram nail.	[61]

Belgian pharmacologist and pharmaceutical chemist Paul Janssen created fentanyl in 1960 [65]. It is a synthetic, short-acting agonist of the -opioid receptor. 50 to 100 times more powerful than morphine is fentanyl. Various new fentanyl analogues started to develop, with furanyl fentanyl, -hydroxythiofentanyl, acryl fentanyl, carfentanil, and valeryl fentanyl being some of the most common. The trend in the usage of fentanyl analogues started with the appearance of -methylfentanyl, followed by acetyl fentanyl. Fentanyl analogue addiction and rising availability have become serious threats to public health globally in recent years. In addition, since the first reports of fentanyl usage deaths more than 30 years ago, there have been hundreds of fentanyl-related fatalities. In recent years, a number of deaths have been reported due to this analog. The UHPLC-MS/MS system utilised by Nan et al. combined an LC system with a triple quadrupole linear ion trap mass spectrometer connected with a TurboIon Spray interface. The separation of the fentanyl analytes took place on a T3 column (100 mm x 2.1 mm x 1.8 m). 20 mmol/L ammonium acetate, 5% acetonitrile, and 0.1 percent formic acid were the components of mobile phase A (MPA). Acetonitrile was used as mobile phase B (MPB). The gradient was set up as follows: 10% B for the first minute, then 90% B for the next six minutes. The technique uses a quick, easy, and affordable liquid-liquid extraction (LLE) to prepare samples, and it only requires a small amount of whole blood (100 l) and runs quickly (8 minutes). Lower limits of quantitation (LLOQs) were in the range of 0.05 to 0.2 ng/ml, while the limits of detection (LODs) were from 0.005 to 0.03 ng/ml. For fentanyl, norfentanyl, and norcarfentanil, the approach was demonstrated to be linear throughout a concentration range of 0.2–40 ng/ml, and 0.05–40 ng/ml for all other target analytes. Recovery rates varied from 72.09 percent to 103.22 percent, while matrix effects ranged from 67.95 percent to 113.32 percent. Additionally, it was discovered that the technique might be used to identify and measure fentanyl and its equivalents in whole blood from actual forensic cases. It was also concluded that this method by Nan et al. has the potential for determining fentanyl group in the real world [58] (Table 2).

Tryptamines are made up of an indole ring with a two-carbon side chain and are derived from the alteration of the amino acid tryptophan [66]. In recent years, hallucinogenic tryptamines have been used as recreational drugs in China. Tryptamines typically cause sensations of euphoria, disinhibition, and aural and visual hallucinations. When users are extremely drunk, tryptamines can regularly cause serious adverse responses, such as tachycardia, tachypnea, hypertension, and hyperthermia. To evaluate the hair of potential users [67], Shi et al. built LC-MS/MS using an Acquity TM Ultra Performance LC ultrahigh pressure liquid chromatography system to concurrently quantify 16 tryptamines in hair. A Waters Acquity UPLC HSS T3 column (100 mm 2.1 mm, 1.8 m) was used for the chromatography, with linear gradient elution occurring at a flow rate of 0.3 mL/min and solvent A consisting of 20 mmol/L ammonium acetate, 5 percent acetonitrile, and 0.1 percent formic acid in water (solvent A) (solvent B). Limits of quantification were from 3 to 50 pg/mg, and limits of detection were between 0.1 and 20 pg/mg. With intraday precision RSDs of less than 14 percent and interday precision RSDs between 1.3 percent and 14 percent, accuracy ranged from 91 to 114 percent. All tryptamine recoveries fell between 85 and 115 percent, with the matrix effect falling between 95 and 112 percent. 191 hair samples from suspected tryptamine users were successfully analyzed using the validated technique; 77 of these samples tested positive for 5-MeO-DiPT, whereas the remaining 114 samples did not contain any of the 16 tryptamines or their metabolites. 34 hair samples included 5-MeO-DiPT as well as its metabolites, 5-MeONiPT, 5-OH-DiPT, and 4-OH-DiPT. The parent chemical discovered in the hair samples was 5-MeO-DiPT [59] (Table 2).

A short-chain fatty acid called gamma-hydroxybutyric acid (GHB) serves as a precursor and metabolite of the inhibitory neurotransmitter gamma-aminobutyric acid (GABA) [68]. In certain nations, sodium oxybate, the sodium salt of this drug, has been licensed and registered as a medicinal agent for the treatment of cataplexy linked to narcolepsy, while in other nations, it is used as an adjuvant drug to aid in the detoxification and withdrawal of alcoholics [69]. GHB has been sold illegally as a recreational drug for the past several years due to its capacity to enhance libido and elicit euphoric sensations. Since it is quickly converted to GHB in the body, its pro-drug gamma-butyrolactone (GBL) has comparable physiological effects when consumed. Measurements of GHB and its metabolites in plasma and urine have been employed in pharmacokinetics investigations on patients treated with GHB for the aforementioned stated disorders to evaluate treatment compliance and confirm potential adverse effects. Utilizing a triple quadrupole mass spectrometer and a Waters Acquity UPLC, UHPLC-MS/MS studies were carried out. An Acquity UPLC HSS reversed-phase C18 column (2.1 mm 150 mm, 1.8 m) was used for the chromatography. Solvent A (0.1 percent formic acid in water) and Solvent B combined to create the mobile phase (methanol). Under isocratic circumstances, 95 percent solvent A and 5 percent solvent B was utilized at a flow rate of 0.3 mL min-1. From the limit of quantification (LOQ) to 500 g mL^{-1} plasma, urine, and cerebrospinal fluid, from LOQ to 100 ng mg-1 hair, and from LOQ to 10 mg mL^{-1} illegal formulations, the approach was linear for all analytes under examination. Recovery of the under-investigation analytes was consistently greater than 75%, and intra- and inter-assay precision and accuracy were greater than 15%. The validated procedure was then successfully used with genuine specimens from forensic (one post-mortem urine sample from a GHB fatal intoxication case) or clinical instances (cerebrospinal fluid, plasma, and hair samples from narcoleptic patients receiving sodium oxybate therapy). Last but not least, police forces' confiscated illegal preparations were examined for GHB content and potential GBL presence [60] (Table 2).

A technique for detecting cannabis in oral fluid using UHPLC-MS/MS and its application to suspected drivers has been developed by Anzillotti et al. THC is the most widely used illicit substance found in drivers. Because of an unfavorable plasma/OF distribution ratio, its concentration in OF might be relatively low. As a result, the suggested cut-off for THC is 2 or even 1 ng/mL in the worldwide standards for OF analysis and significant research conducted on DUID. For evaluating drugged driving, the confirmation of 9-tetrahydro-cannabinol (THC) in oral fluid (OF) is crucial (Driving Under the Influence of Drugs, DUID). The importance of creating such analytical techniques, as this researcher outlines, is of particular interest to DUID, especially in light of the various and constantly evolving legal policies regarding marijuana use and possession around the world, such as the state-level drug laws in the United States. The detection system was an Agilent 6460 triple quadrupole mass spectrometer with the Jet Stream electrospray ionisation source. The UHPLC system was an Agilent 1290 Infinity system with binary pump and integrated vacuum degasser. The column was a 2.6 m, 100 2.1 mm Kinetex C18 column. The lowest limit of detection (LLOD) for THC in UHPLC-MS/MS and SPME-GC/MS was 2 ng/mL and 0.5 ng/mL, respectively.

Additionally, the LLOD for cannabidiol (CBD) and cannabinol (CBN) in GC/MS equipment was 2 ng/mL, whereas it was 20 ng/mL in UHPLC-MS/MS. The developed method was found to be more suitable for precision, accuracy and linearity [61]. Both keratinized biological matrices—hair and nails—can absorb psychoactive substances and keep them locked up for an extended period. Hair analysis has long been recognized as a valuable tool in toxicological investigations where evidence of repeated drug use in clinical and forensic contexts is required, such as assessment of chronic psychoactive drug administration, drug misuse monitoring, workplace drug testing, and evaluation of driving license renewal. More recently, nails have been suggested and effectively used as a replacement or addition to hair in fatal instances or drug-related intoxications (Table 2).

Mannocchi et al. goal was to create and thoroughly verify a UHPLC-MS/MS screening technology to quantify 119 molecules among the most commonly abused traditional pharmaceuticals and novel psychoactive compounds in hair and nails to gauge polydrug usage. The standard internal combination was added to 25 milligrams of cut hair or nail samples, which were then incubated at a regulated temperature in 500 L of buffer reagent. After cooling, a chromatographic system with an Oasis HLB column was used, and 1 L of supernatant was added. The target compounds were discovered in multiple reaction monitoring modes following the 10-minute chromatographic separation through a mobile gradient phase (aqueous ammonium formate, phase A; acetonitrile, phase B). The investigational analytes' process efficiency was consistently more than 65%, and no discernible ion suppression from the matrix effect was noticed. Precision and accuracy within and between assays were consistently higher than 15%. Analyzing actual hair and nail samples from people who take psychoactive substances for recreational purposes allowed researchers to test the method's applicability and veracity. It is possible to identify both traditional drugs and novel psychoactive chemicals due to single or recurrent usage and buildup in keratin matrices [62] (Table 2).

3.3. Implementation of UHPLC-MS/MS in Metabolite Analysis

Cannabis sativa L. and its subspecies Cannabis indica L. and Cannabis ruderalis L. all include a group of chemical compounds called phyto-cannabinoids. Due to their pharmacological effects, -9-tetrahydrocannabinol (THC) and cannabidiol (CBD) are the most prevalent and researched phytocannabinoids [76]. For example, although CBD exhibits antiepileptic, anxiolytic, and neuroprotective characteristics without inducing psychotropic or addictive effects, THC exhibits psychotropic and addictive impact and is a partial agonist at CB1 and CB2 receptors [77]. CBD is an antagonist at CB1 and CB2 receptors. Pichini *et al.* developed a method for quantifying cannabidiol (CBD) and its metabolites, cannabidiol-7-oic acid (7-COOH-CBD), 7-hydroxycannabidiol (7-OH-CBD), 6-alpha-hydroxycannabidiol (6-OH-CBD), and 6-beta hydroxycannabidiol (6-OH-CBD), in serum and urine samples of a person treated with medical cannabis was developed and validated using UHP To achieve the sensitivity needed to identify trace levels, the ionisation was carried out using an electrospray in negative mode, with limits of quantification ranging from 0.05 to 0.1 ng/mL. The approach is quick, accurate (inter/intra-day error average of 15%), and exact (inter/intra-day imprecision average of 15%). (8 min run time). The technique is a crucial tool for examining the pharmacokinetics and pharmacodynamics of CBD in patients receiving medicinal marijuana or CBD-based pharmaceuticals [70] (Table 3).

For more than 40 years, the C8-branched short-chain fatty acid valproic acid (VPA, 2-propylvaleric acid) has been utilized as a broad-spectrum antiepileptic medication against many forms of seizures and epileptic disorders [78]. VPA has a variety of modes of action and an acceptable safety profile, making it a standard treatment for paediatric epilepsy. Despite its undeniable pharmacological relevance and efficacy, VPA's possible hepatotoxicity is still a severe worry. A boxed warning for potentially fatal adverse medication reactions, such as pancreatitis and hepatotoxicity, is present on the drug label [79]. Zhao et al. employed an Agilent 1290 system, which includes a binary pump, an autosampler, and a temperature-controlled column compartment, with a RHD EC-C18 column (1.8 m, 2.1 mm 75 mm). To separate the analytes, the mobile phase was composed of acetonitrile (A) and 10 mM ammonium acetate (B) at a flow rate of 0.6 mL/min. An electrospray ionization interface was used to link the UHPLC equipment to the AB SCIEX 4500 QTRAP mass spectrometer. The acceptable standards of 15 percent for inter-day and intra-day accuracy and precision were discovered to be met. The recoveries and matrix effects satisfied the requirements for analyzing biological samples. Under varied storage circumstances, including room temperature for 12 hours, three freeze-thaw cycles, and 80°C for one month, no noticeable deterioration was seen. The test technique successfully tracks the levels of VPA and its three metabolites in individuals with epilepsy. The UHPLC-MS/MS approach showed a robust analytical capability necessary for therapeutic drug monitoring, which may result in clinically significant advancements in VPA dose and patient care [71] (Table 3).

Potential biomarkers for cancer diagnosis and therapy assessment include polyamines and their N-acetylated metabolites. To simultaneously determine 6 free, 3 monoacetylated, and 2 diacetylated polyamines without derivatization, we describe an ultra-high-performance liquid chromatography-tandem mass spectrometry (UHPLCMS/ MS) approach. Yu and coworkers utilised on an XR LC-20AD Prominence TM UFLC system with a SHIMPACK XR-ODS column (75 x 3.0 x 2.2 mm), liquid chromatography separation was performed. 0.0125 percent PFPA in water (A) and 0.0125 percent PFPA in methanol make up the mobile phase (B). A Turbo Ion Spray

Table 3. Applications of UHPLC-MS/MS in metabolite analysis

S. No	Analytes	Matrix	Condition	Run time	R2 value	LOD	LLOQ	Ref.
1	Cannabidiol metabolties	Serum and Urine sample	BEH C18 column (100 mm, 2.1 mm, 1.7 µm), 0.4 mL/min,	8 min	0.998 0.997 0.998 0.999	0.02 0.02 0.02 0.02	0.05 0.05 0.05 0.05	[70]
					0.997 0.998 0.997 0.998	0.02 0.02 0.02 0.02	0.05 0.05 0.05 0.05	
2	Valproic acid and its metabolites	Human serum	C18 column (1.8 µm, 2.1 mm × 75 mm), 0.6 mL/min, 35°C		0.9956 0.9964 0.9984 0.9982		1 µg/mL 0.5 µg/mL 10 ng/mL 2.5 ng/mL	[71]
3	Free polyamines and their metabolites	Plasma, Urine and Liver tissue	XR-ODS column (75×3.0 mm, 2.2 µm),		0.9904-0.9986 in plasma 0.9912 -0.9985 in urine 0.9912 - 0.9983 in liver tissue			[72]
4	20 mycotoxis and major metabolits	Water Sample	Acquity HSS T3 100 × 2.1 mm 1.8 µm), 0.3 mL min-1, 40°C		≥ 0.9908	0.04 to 27.91 ng L^{-1}	1 and 40 ng L^{-1}	[73]
5	Metabolites of Aspirin	Rat Urine	Acquity BEH C18 column (2.1 mm i.d. × 100 mm, 1.7 µm), 0.4 ml/min, 35°C	7 min	0.9736 0.9549 0.9943		0.67µg/mL 0.28 µg/mL 0.25 µg/mL	[74]
6	17 isoprostanes, isomeric prostanoid mediators, three nitro fatty acids, four sphingoid mediators, and 24 lysophosphatidic acid species	Human Serum			0.997 to 0.9999	0.09-1nM		[75]

source-equipped QTRAPTM 4000 MS/MS system from Applied AB Sciex was used for MS detection. Electrospray positive ionisation (ESI+) was used for the analytes detection's multiple reaction monitoring modes (MRM). According to the method's conclusion, hepatocellular carcinoma may now be diagnosed and treated in a way that is more flexible in terms of practical applicability [72] (Table 3).

Secondary metabolites known as mycotoxins are produced by filamentous fungus when they grow on organic matter. Since many mycotoxins have poisonous qualities and regularly contaminate food and feed items from crops, such as grains, maize, nuts, seeds, fruits, beer, and wine, they are known for their potential hazard to human and animal health [80]. More than 100 identified filamentous fungus produce a wide range of mycotoxins with various structures, chemical makeups, and physical makeups, leading to different hazardous effects on animals, plants, and microbes. Gradient elution for the positive ionization mode was created with a mobile phase consisting of 0.01 percent formic acid in ultrapure water (mobile phase A) and MeOH containing 1 mM ammonium formate (mobile phase B) in the study by Goessens et al. A Waters Xevo® TQ-XS triple quadrupole mass spectrometer with an electrospray ionisation (ESI) interface was used to identify the analytes. Further investigation into the dynamics and ecotoxicological effects of mycotoxins in aquatic environments was shown to be aided by this strategy [73] (Table 3).

One of the most common freshwater-farmed fish species is tilapia, a prominent species in tropical and subtropical parts of the world. Fish is frequently processed thermally, including boiling, steaming, and air frying [81]. Lipid oxidation and the Maillard reaction are important processes in fish flesh processing that provide volatile flavors. However, several non-volatile ingredients in raw and cooked meats, such as lipids, vitamins, nucleosides, amino acids, and organic acids, are also vital taste precursors. Li et al. employed the following scenario for their exploratory analysis, which involved using a 2.1 mm 100 mm ACQUIY UPLC BEH 1.7 m column. The mobile phase contains A = 25 mM ammonium acetate and 25 mM ammonium hydroxide in water, and B = acetonitrile in both ESI positive and negative modes. The AB Sciex Triple TOF 6600 mass spectrometer was used for mass spectrometry analysis. Positive and negative ESI (Electrospray Ionization) techniques were employed for detection. The findings demonstrated that thermally processed tilapia fillets might be distinguished using untargeted metabolomics based on UHPLC-Q-TOF-MS/MS. The method's usefulness for examining changes in the metabolic makeup of aquatic products after thermal processing is further shown by this study [74] (Table 3).

A common analgesic and anti-inflammatory drug, acetylsalicylic acid is used at dosages daily more than 325 mg [82]. Currently, ASA is frequently used for cardiovascular disease at modest levels (325 mg per day, 75–100 mg daily in most people). The work described here aims to develop a quick and accurate approach to measure ASA metabolites in biological matrices, particularly for glucuronide metabolites for which commercial standards are not yet available. To separate the analytes, Ping Li et al. employed an Acquity UHPLC BEH C18 column (2.1 mm i.d. 100 mm, 1.7 m). The mobile phase was made up of (A) formic acid-water (1:1000, v/v) and (B) acetonitrile. With this technique, it is possible to assess how much ASA impacts the metabolic and excretion profile of ASA metabolites in rat urine. Salicyl phenolic glucuronide (SAPG) and salicyluric acid phenolic glucuronide (SUAPG) was measured indirectly by measuring the liberated salicylic acid (SA), gentisic acid (GA), and 4 salicyluric acid (SUA) from SAPG and SUAPG following - glucuronidase digestion. 24 hours after ASA treatment, the primary metabolites of ASA in rat urine were SUA and SUAPG, accounting for 50% (SUA) and 26% (SUAPG), respectively (SUAPG). As ASA dose was raised, the contributions decreased to 32 percent and 18 percent. However, the excretion of the other three metabolites (GA, SA, and SAPG) significantly increased by 16%, 6%, and 4%, respectively. The study showed that ASA dose might considerably impact the metabolite metabolism and excretion profile of ASA in rat urine [75] (Table 3).

An underlying pathogenic mechanism linked to most illnesses is oxidative stress, defined as a state where the amounts of reactive oxygen species (ROS) required for cellular redox biology rise over the cellular antioxidant threshold, causing macromolecular damage. A Kromasil eternityXT-1.8 C18 column (50 mm x 2.1 mm, 1.8 m) was utilized by Schoeman et al. Water, 5 mM ammonium acetate, 0.0625 percent ammonium hydroxide, and ACN-IPA (8:2 v/v) made up mobile phase A, whereas 0.1 percent ammonium hydroxide made up mobile phase B. The Shimadzu LCMS-8050 system comprises a triple-quadrupole mass spectrometer with an electrospray ionization source linked to a UHPLC system. It outlines a quick, sensitive, and focused UHPLC-MS/MS metabolomics technique that enables the biological elucidation and quantitative measurement of 17 isoprostanes, their corresponding isomeric prostanoid mediators, three nitro fatty acids, four sphingoid mediators, and 24 lysophosphatidic acid species from serum as well as organ tissues, such as liver, lung, heart, spleen, kidney, and brain. The application of this technique on serum and tissue samples from paired mice revealed stress and inflammatory readouts unique to the tissue and serum. In a homeostatic model, there was little association between localised (tissue) metabolite levels and systemic (serum) circulation. It was determined that this approach would allow us to investigate the function of signalling lipids in the metabolic pathogenicity of stress and inflammation during health and illness in future investigations [76] (Table 3).

3.4. Method Development Using UHPLC-MS/MS Technique

Triterpenoid chemicals, such as oleanolic acid (OA) and ursolic acid (UA), are often found in plants and fruits. Important pharmacological qualities, such as anti-inflammatory, hepatoprotective, anti-ulcer, anti-microbial, anti-tumor, anti-HIV, antihyperlipidemic activity, and others, have been documented for them. Zheng et al. used the Agilent 1290 UHPLC system with the Agilent 6460 Triple Quadrupole MS/MS system made formed the UHPLC-MS/MS system. The chromatographic separation was carried out on an Agilent SB C18 column (2.1 mm 50 mm, 1.8 m). Eluent B was acetonitrile and included 0.1 percent formic acid, while Eluent A had 5 percent acetonitrile/water. The detection limits for oleanolic acid and ursolic acid were 0.025 and 0.020 ng/mL, respectively. The validated approach was found to be promising for the sensitive, exact, and simultaneous measurement of oleanolic acid and ursolic acid [83] (Table 4).

Neonicotinoids are among the most frequently used insecticides for crop protection in agriculture [94]. Neonicotinoids are transported to all plant organs after administration as a seed treatment, soil application, or foliar spraying and can therefore shield plants against harm by herbivorous insects [95]. However, in recent years there has been an increase in worry over neonicotinoids' effects on non-target animals, particularly insect pollinators like bees. These pesticides are present in pollen and nectar, which are some pollinators' only food sources because of their systemic method of action. Neonicotinoids are also highly stable, soluble in water, and therefore have the potential to stay in the environment. As a result, they may pose a threat to the environment and a security risk to human health due to the consumption of tainted food. Lachat et al. developed a method to quantify the most popular neonicotinoids (thiamethoxam, clothianidin, imidacloprid, acetamiprid, and thiacloprid) at trace levels in milk. The method was developed and validated

Table 4. Method development for analytes using UHPLC-MS/MS

S. No	Analytes	Matrix	Condition	Run time	R2 value	LOD	LLOQ	Ref.
1	Oleanolic acid Ursolic acid	Rat blood	Agilent SB C18 column (2.1 mm × 50 mm, 1.80 μm), 2 mL/min, 30°C		0.991 0.995	0.025 ng/mL 0.020 ng/mL	0.090 ng/mL 0.080 ng/mL	[83]
2	Neonicotinoids	Milk	50 mm x 2.1 mm i.d., 1.6 μm, Cortecs UPLC C18 column, 0.4 mL/min, 25°C.	8.50 min	0.999		2 pg/mL 5 pg/mL 2 pg/mL 0.5 pg/mL 1 pg/mL	[84]
3	12 constituents from *Erigeron breviscapus*	Rat Plasma	Eclipse Plus C18 column (2.1 mm × 50 mm, 1.8 μm), 0.3 mL/min^{-1}, RT		0.9991 0.9995 0.9995 0.9992 0.9991 0.9992 0.9990 0.9993 0.9991 0.9990 0.9995 0.9996	8.0 ng/mL 1.6 ng/mL 50 ng/mL 2.0 ng/mL 0.06 ng/mL 0.08 ng/mL 50 ng/mL 0.08 ng/mL 16 ng/mL 10 ng/mL 6.0 ng/mL 1.0 ng/mL	20 ng/mL 3.0 ng/mL 100 ng/mL 7.0 ng/mL 0.3 ng/mL 2.0 ng/mL 70 ng/mL 1.0 ng/mL 20 ng/mL 30 ng/mL 10 ng/mL 2.0 ng/mL	[85]
4	Kaempferol and its metabolite 4-hydroxyphenyl acetic acid	Caco-2 Cells and immortalized monoculture human model and a primary triple-co-culture rat model	Eclipse Plus C18 column (2.1 × 50 mm, 1.8 μm), 0.4 ml/min, 55°C		0.995 0.997			[86]

S. No	Analytes	Matrix	Condition	Run time	R2 value	LOD	LLOQ	Ref.
5	Phenacetin, omeprazole, metoprolol, Midazolam and their metabolites	Rat Plasma	Eclipse Plus C18 column (1.8 μm, 2.1×50 mm, I.D), 0.4 mL/min, 40°C	3.0 min	0.9952 0.9949 0.9957 0.9949 0.9985 0.9987 0.9920		1 ng/mL, 10 ng/ml	[87]
6	curcumin, tetrahydro-curcumin, quercetin, and paeoniflorin	Rat Plasma	C18 column (50 mm×2.1 mm, particle size 3 μm), 0.3 mL/min, 25℃.	3.0 min			1.0 ng/mL, 1.0 ng/mL, 2.0 ng/mL, 10.0 ng/mL	[88]
7	Gut-derived uremic toxin	Human Serum	BEH C18 (2.1 × 100 mm, 1.7 μm) column, 0.3 mL/min, 35°C	4 min	≥0.997			[89]
8	Phenolic compounds in Chamomile and Chamomile tea extracts		BEH C18 (100 × 2.1 mm, 1.7 μm), 0.45 ml/min.,		> 0.9982	0.002-0.015 μmol/L	0.005-0.050 μmol/L	[90]
9	17 steroidal hormones	Equine serum	RRHD C18 analytical column (50 x 2.10 mm i.d., 1.8 μm), 0.6 ml/min,	11 min	0.9922–0.9986	0.002 ng/mL 0.03 ng/mL 0.2 ng/mL 0.02 ng/mL 0.2 ng/mL 0.2 ng/mL 0.2 ng/mL 0.3 ng/mL 0.5 ng/mL 0.02 ng/mL 0.025 ng/mL	0.0055 ng/mL 0.055 ng/mL 0.55 ng/mL 0.05 ng/mL 0.55 ng/mL 0.55 ng/mL 0.55 ng/mL 1.1 ng/mL 1.1 ng/mL 0.05 ng/mL 0.05 ng/mL	[91]

Table 4. (Continued)

S. No	Analytes	Matrix	Condition	Run time	R2 value	LOD	LLOQ	Ref.
9 (Continued)						0.05 ng/mL 0.02 ng/mL 0.05 ng/mL 0.02 ng/mL 0.2 ng/mL 2 ng/mL	0.11 ng/mL 0.055 ng/mL 0.11 ng/mL 0.055 ng/mL 0.55 ng/mL 5.5 ng/mL	
10	Hop Prenyl-flavonoids	Human Serum	C18 column (2.0 × 50 mm, 1.6 µm). 0.5 mL/min.	2.5 min	>0.997		0.50 ng/mL 1 ng/mL	[92]
11	Bisphenol A and its chlorinated derivatives	Human Plasma	CSH C18 (1.7 µm particle size, 2.1 x 100 mm), 350 $\mu L.min^{-1}$,		0.996 0.997 0.998 0.997 0.990		0.05 ng/mL	[93]

using UHPLC-MS/MS technique. Neonicotinoids were separated on a Cortecs UPLC C18 column (50 mm 146 x 2.1 mm i.d., 1.6 m). Mobile phase A was acetonitrile with 0.05 percent formic acid; mobile phase B was water with 0.05 percent formic acid and 5 mM ammonium formate. The technique assessed neonicotinoid concentrations in Swiss milk, including ordinary and organic milk, human breast milk, and commercial powdered milk. In more than 90% of the samples, at least one neonicotinoid was present. However, with average total neonicotinoid levels of 16.1 13.1 pg/mL, all animal samples were well below the maximum residual limits permitted for human consumption. Neoicotinoids were present in human breast milk and powdered milk at comparable levels. When seen collectively, our findings show the substantial frequency of neonicotinoids in milk from all sources, albeit at concentrations deemed acceptable for ingestion by people [84] (Table 4).A significant traditional Chinese herbal is *Erigeron breviscapus* (Vant.) Hand-Mazz (Compositae) as a complete plant. *Erigeron breviscapus*, which possesses the effect of anticoagulant, antifungal, antibacterial, neuroprotective, and visual field protective activities, is widely used for the therapy of stroke hemiplegia, kidney failure, angina pectoris, coronary disease, pulmonary heart disease, diabetes and cardio cerebral diseases. The twelve primary bioactive ingredients (neochlorogenic acid, chlorogenic acid, caffeic acid, cynarin, scopoletin, scutellarin, isochlorogenic acid A, apigenin-7-o-glucuronide, isochlorogenic acid C, scutellarein, luteolin, and apigenin) in rat plasma were simultaneously determined using a method developed and validated by Tian et al. On an Agilent ZORBAX RRHD Eclipse Plus C18 column (2.1 mm 50 mm, 1.8 m), chromatographic separation was carried out using a gradient of 0.1 percent formic acid aqueous solution and acetonitrile as the mobile phase. The lower limits of quantification (LLOQ) for neochlorogenic acid, chlorogenic acid, caffeic acid, cynarin, scopoletin, scutellarin, isochlorogenic acid A, apigenin-7-o-glucuronide, isochlorogenic acid C, scutellarein, luteolin, and apigenin were 20, 3, 0, 100, 7.0, 0.30, 2.0, 70, 1.0, 20, 30, 10. The needed limitations were discovered to be stability, matrix influence, and extraction recovery. The pharmacokinetics of twelve components after oral administration to research study concerning *Erigeron breviscapus* of clinical practices for individual analytes on rats were investigated using this approach, which was both sensitive and selective [85] (Table 4).

Pharmacological investigations have shown that flavonoids, including kaempferol and quercetin, as well as several of its intestinal metabolites, exhibit sedative and anxiolytic-like actions. Routes of delivery, however, were found to be essential for monitoring *in vivo* activity.

Kaempferol (KMF) and its primary intestinal metabolite, 4-hydroxy-phenylacetic acid, were tested for their capacity to pass intestinal and blood-brain barriers in cell-based models (4-HPAA). Caco-2 cells were used for intestinal transport investigations, while immortalised monoculture human and primary triple-coculture rat models were used for blood-brain barrier transport research. The methodologies were employed for *in vitro* research in well-known models for intestinal and BBB permeability after they had undergone validation by EMA and FDA requirements [86] (Table 4).

Cytochrome P450 activities are hypothesized to be inhibited during drug-drug interactions (DDIs) [96]. To ascertain CYP450 activity, the cocktail approach includes investigating the metabolism of two or more probe medicines. For the simultaneous quantification of four CYP450 probe drugs—phenacetin, omeprazole, metoprolol, and midazolam—and their metabolites—acetaminophen, 5'-hydroxy omeprazole, -hydroxy metoprolol, and 1'-hydroxy midazolam—in rat plasma, Wang et al. developed a UHPLC-MS/MS technique. In about 3.0 minutes, the separation was completed on a ZORBAX Eclipse Plus C18 Rapid Resolution High Definition column utilizing gradient elution and water containing 0.1 percent formic acid (A) and acetonitrile (B). A 6420 series triple-quadrupole tandem mass spectrometer was used for detection, employing positive ion mode ESI with multiple reaction monitoring (MRM). For phenacetin, omeprazole, metoprolol, and midazolam, as well as 1-500 ng/mL for their metabolites, the calibration curves were linear across the concentration range of 10-5000 ng/mL. Precisions within one day and between days were within 15 percent, and accuracies ranged from 87 to 112 percent. After administering a single oral dosage of phenacetin, the approach was used to successfully analyse the pharmacokinetics of probe pharmaceuticals, metabolites, and DDIs using 3-n-butylphthalide (NBP) [87] (Table 4).

Along with quercetin (QR) and paeoniflorin, curcumin (CUR) is a bioactive component included in several composite traditional Chinese medicine prescriptions (PF). When all three drugs are taken orally, little is known about how QR and PF affect CUR's absorption and metabolism. Yu et al. in their studies, Used tinidazole as the internal standard. A quick, sensitive, and trustworthy ultra-high-performance liquid chromatography-tandem mass spectrometry (UHPLC-MS/MS) approach was created and validated for the simultaneous detection of CUR, tetrahydrocurcumin (THC), QR, and PF in rat plasma (IS). The plasma samples were pre-treated by ethyl acetate liquid-liquid extraction. Acetonitrile and 0.1 percent formic acid water solution (80:20, v/v) were used as the mobile phase during chromatographic separation, carried out on a C18 column with isocratic elution at a flow rate of 0.3 mL/min. To identify m/z transitions, a TSQ Quantum Access Max API mass spectrometer with an electrospray ionisation (ESI) source in selection reaction monitoring (SRM) mode was used. The method's stability, matrix effect, extraction recovery, precision, selectivity, and accuracy were all verified. This established and verified methodology was effectively used in the pharmacokinetic research of CUR, THC, QR, and PF in rats [88] (Table 4).

In individuals with underlying chronic renal illness, gut-derived uremic toxins are becoming more widely recognized as potentially modifiable cardiovascular disease risk factors contributing to uremic syndrome. An ultra-performance liquid chromatography-tandem mass spectrometry approach was designed and validated to simultaneously measure a panel of four gut-derived uremic toxins in human serum. The panel comprises kynurenic acid, hippuric acid, indoxyl sulphate, and p-cresol sulphate. With deuterated internal standards and acetonitrile, serum samples were protein precipitated. An Acquity BEH C18 (2.1 100 mm, 1.7 m) column was used for the chromatographic separation of the analytes, and isocratic elution at a flow rate of 0.3 mL/min was used with a mobile phase made up of solvent A (10 mM ammonium formate; pH 4.3) and solvent B (acetonitrile) (85:15, v/v). Utilizing hot electrospray ionisation, analyses were found. For the LLOQs and all other quality controls, intra- and inter-day accuracy and precision were within 19.3% and 10.9% of each other, respectively. Matrix impact from serum was 15%, and recovery for all analytes was 81.3%. The technique uses a quick run-time, straightforward, low-cost sample processing, has passed FDA validation guidelines and has been effectively used to research renal disease patients by Prokopienko et al. [89] (Table 4).

The medicinal herb chamomile (*Matricaria recutita L.*) has sedative, analgesic, anti-allergic, anti-spasmodic, antibacterial, anti-inflammatory, and anti-spasmodic effects [97]. The most often utilized form of it is its essential oil, which contains volatile chemicals and terpenoids such azulene, chamazulene, and α-bisabolol. Recently, the non-volatile portion of the extract has also received attention, particularly for the antioxidant and spasmolytic properties of phenolic chemicals. Chamomile was shown to have 0.42 quercetin equivalent or 1.30 trolox equivalent antioxidant activity. Novakova et al. used gradient elution to achieve the separation, with methanol and formic acid at 0.1 percent as the mobile phase. To achieve excellent sensitivity and selectivity for quantification utilising the SRM experiment, ESI in both positive and negative ion mode was tuned. Except for chlorogenic acid and kaempferol, which showed higher findings of linearity, accuracy, and precision in ESI positive ion mode, it was discovered that ESI in negative ion mode was more practical for quantitative analysis of all phenolics. The method validation results confirmed that the developed UHPLC-MS/MS method was practical and trustworthy for determining phenolic compounds in Chamomile extracts. The usual range for method sensitivity indicated as LOQ was 5–20 nmol/L [90] (Table 4).

A huge group of substances derived from cholesterol called steroids are essential for the transmission of a wide variety of biological signals inside the body. Their general duties may be divided into four categories: management of the internal environment, development and growth, reproduction and sexual differentiation, and regulation of metabolism and nutrition supply. In contrast to human sports, controlling doping in equestrian competitions presents unique difficulties. Using the UHPLC-MS/MS technique, Genangeli et al. have created a new, quick, and straightforward analytical method that can simultaneously identify and quantify 17 steroid hormones and metabolites (Pregnenolone, 17-OH-Pregnenolone, Progesterone, 17-OH Progesterone, Androsterone, Androstenedione, DHEA, DHEAS, Testosterone, Cortisol, CortiCorticosterone, Aldosterone, 11-Deoxycortisol, 11-Deoxycorticosterone, Dihydrotestosterone, Estrone, Estradiol). An Agilent 1290 Infinity series and an Agilent Technology Triple Quadrupole 6420 equipped with an ESI source operating in negative and positive ionisation modes were used to conduct UHPLC-ESI-MS/MS investigations. A Zorbax RRHD C18 analytical column (50 x 2.10 mm i.d.1.8 μm) was used to separate the analytes. A combination of water (A) and acetonitrile (B) both containing 0.1 percent formic acid served as the mobile phase for UHPLC-MS/MS studies. Following validation, the technique was applied to measure 17 steroid hormones in blood samples from mare, stallion, and geldings. Cortisol (range 32.57-52.24 ng ml-1) and corticosterone (range 37.25-51.26 ng ml-1) were the primary steroids found, followed by 17-OH-pregnenolone, dihydrotestosterone, and pregnenolone [91] (Table 4).

In the brewing business, hops (*Humulus lupulus L.*) are used as a flavor and preservative for beer. Due to their wide range of biological actions, prenylated polyphenols have garnered the greatest attention among the active components of hops. Hop cones (female inflorescences) contain large amounts of xanthohumol (XN), a prenylated chalcone being studied for its potential as a cancer chemopreventive. Preparations of hops that include the estrogenic prenylflavonoids 8-prenylnaringenin (8-PN), 6-prenylnaringenin (6-PN), and isoxanthohumol (IX) are being studied as alternatives to conventional hormone replacement therapy for the treatment of menopausal symptoms in women. Yuan et al. created a straightforward procedure that involves enzymatic hydrolysis, liquid-liquid extraction, and 2.5 minutes of chromatographic separation using a reversed-phase C18 column. A triple quadrupole mass spectrometer outfitted with electrospray ionisation was used for the analyte detection step (Table 4).

In conclusion, a technique for measuring the levels of four-hop prenylflavonoids in human blood was created and validated using UHPLC-MS-MS. UHPLC-MS-MS significantly boosts analytical speed and sensitivity compared to other current methods while retaining high accuracy and precision. The detection of prenylflavonoids in the serum of human patients from an ongoing phase I clinical study was determined to be appropriate for the novel UHPLC-MS-MS technology [92].

A well-known chemical called bisphenol A (BPA), which is present in resins including polycarbonate, polysulfone, and epoxy, is employed in the mass manufacture of numerous consumer goods [98]. Endocrine disrupting qualities of BPA have the potential to cause negative health impacts. In aquatic conditions, it can react with chlorine to create chlorinated derivatives (ClxBPAs). ClxBPAs have oestrogenic activity that is 10 to 105 times more than that of BPA alone. To evaluate health risks, it is essential to evaluate human exposure to endocrine disruptive substances. Human biomonitoring frequently makes use of blood and urine matrix. A method based on Ultra High-Performance Liquid Chromatography coupled to a Triple Quad Mass Spectrometer was developed, fully validated, and used by Cambien et al. to identify BPA, monochlorobisphenol A (MCBPA), dichlorobisphenol A (DCBPA), trichlorobisphenol A (TCBPA), and tetrachlorobisphenol A (TTCBPA) in human blood plasma. They utilised Acquity CSH C18 (1.7 m particle size, 2.1 x 100 mm), and a binary mobile phase was provided in gradient mode with the help of the solvents Water (A) and MeOH (B). An API 6500+ mass spectrometer running in negative ionization mode and with an electrospray ionisation (ESI) interface was used for the quantification. A method based on Ultra High Performance Liquid Chromatography coupled to a Triple Quad Mass Spectrometer was developed, fully validated, and used by Cambien et al. to identify BPA, monochlorobisphenol A (MCBPA), dichlorobisphenol A (DCBPA), trichlorobisphenol A (TCBPA), and tetrachlorobisphenol A (TTCBPA) in human blood plasma. They utilized Acquity CSH C18 (1.7 m particle size, 2.1 x 100 mm), and a binary mobile phase was provided in gradient mode with the help of the solvents Water (A) and MeOH (B). An API 6500+ mass spectrometer running in negative ionisation mode and with an electrospray ionisation (ESI) interface was used for the quantification. It was shown that the new approach is quite sensitive for detecting BPA and ClxBPAs in human plasma [93] (Table 4).

3.5. Miscellaneous Applications

Sulfonamides (SAs) are a large class of antibacterial substances often employed in human and veterinary medicine. SAs are amphoteric compounds with high water solubility and low molecular weight [103]. Common anthropogenic contaminants known as SAs have been found in significant Asian and European groundwater, lakes, and rivers. The investigations presented here evaluate the content and dispersion of certain sulfonamide antibiotics in surface water samples taken from a heavily populated area. A Kinetex F5 (2.1 mm; 1.7 m) column was employed by Kokoszka et al. and elution was accomplished using a gradient system with a flow rate of 0.3 mL per min. Formic acid solution diluted in water at 0.1 percent and acetonitrile made up the mobile phase. A tandem mass detector was attached to the chromatographic apparatus previously stated.

The positive ionization mode was used to analyze each of the chosen SAs. The most typical contaminants in surface waters were sulfamethoxazole (maximum 78.88 ng L^{-1}) and sulfapyridine (maximum 38.88 ng L^{-1}). Sulfadiazine was also found in trace levels (below LOQ = 0.40 ng L-1). The samples were subsequently tested to look for transformation products. In the ambient samples, several sulfadiazine and sulfamethoxazole transformation products were found and verified [99] (Table 5).

A genus of rod-shaped, Gram-negative bacteria known as *Salmonella spp.* may be found in the intestinal flora of several animals. [104] It frequently results in food poisoning reports, resulting in gastrointestinal issues. The primary cause of human salmonellas is eating raw or undercooked shell eggs. The ingestion of tainted chicken products is the leading cause of *Salmonella* spp. Infections in humans. Because there is a chance that resistance will emerge and spread from animals to people. The development of antibiotic resistance and the transmission of resistant *Salmonella* spp. bacteria to humans are serious concerns, second only to the danger of sickness. Salmonella species' antibiotic resistance profile is attributed to a gene cluster carried by chromosomal genomic island 1. (SGI1). The determination of antimicrobial residues in poultry excreta using ultra-high performance liquid chromatography-tandem mass spectrometry (UHPLC-MS/MS) was developed and

Table 5. Miscellaneous applications of UHPLC-MS/MS

S. No	Analytes	Matrix	Condition	Run time	R2 value	LOD	LLOQ	Ref
1	Sulfonamide and transformation products	Surface water	A Kinetex F5 (150 × 2.1 mm; 1.7 μm), 0.3 mL/min1, 25°C				0.40 ng /L	[99]
2	Veterinary anti-microbials	Poultry excreta	BEH C18 column (100 mm×2.1 mm i.d., 1.7 μm), 0.4 mL/min, 40°C	7 min	0.996 0.995 0.996 0.999 0.995 0.998 0.998 0.998 0.995 0.996 0.996	7.50 ng/g 3.17 ng/g 9.05 ng/g 2.19 ng/g 7.76 ng/g 1.84 ng/g 8.50 ng/g 9.22 ng/g 8.39 ng/g 7.93 ng/g 2.47 ng/g	50 ng/g	[100]
3	Peptide Exendin-4 (39 amino acids in length)	Dried blood spot (Monkey plasma)	BEH C18, (2.1 × 50 mm, 1.7 μm), 1 ml/min, 70°C		0.9966		20 ng/ml	[101]
4	7 Statins	Serum	BEH C18 (50 × 2.1 mm; 1.7 μm), 0.5 ml/min, 4°C		> 0.997		0.5-5 ng/mL	[102]

validated by Gorissen et al. for classes having an active spectrum against *Salmonella* spp. in poultry:- lactams (amoxicillin and penicillin V), fluoroquinolones (enrofloxacin, difloxacin, and flumequine), polymyxins (col (chlortetracycline and doxycycline). The column utilized was an Acquity C18 (100 mm 2.1 mm i.d., 1.7 μm). Formic acid in aqueous form comprises mobile phase A, whereas formic acid in aqueous form makes up mobile phase B. A XevoTMTQ-S triple quadrupole mass spectrometer with a positive-mode electrospray ionisation (ESI) source was used for the analysis. The described approach was used in Salmonella control programmes because it combines the benefits of a general, straightforward sample preparation procedure with the specificity of UHPLC-MS/MS. Additionally, in terms of animal welfare and labor effort, the noninvasive sample collection of excreta is greatly wanted in the field [100] (Table 5).

Over the past five years, dried blood spot (DBS) analysis has grown in popularity and received much attention in the literature. This technique has been chiefly used for small molecule quantification in the pharmaceutical sector. This study aimed to determine whether the benefits of the 3 R's (reduction, refinement, and replacement) provided by DBS technology might also be conveniently applied to the support of biopharmaceutical studies. The question of whether dried blood spots (DBS) might be used to quantify biopharmaceuticals in biological matrices and operate as accurately, precisely, and steadily as conventional plasma techniques were investigated. A method was successfully validated using DBS technology for the Exendin-4 (39 amino acids in length). Compared to an existing monkey plasma approach, the verified DBS method produced a more sensitive and straightforward process and required a tenfold less sample amount. The final DBS approach produced a linear calibration range of 10-2000 ng/ml using around 5 l of dried blood, whereas the plasma method produced a linear calibration range of 20-2000 ng/ml utilizing 150 μL of plasma and SPE sample preparation [101] (Table 5).

The crucial enzyme for cholesterol manufacture, 3-hydroxy-3-methylglutaryl coenzyme A reductase, is competitively inhibited by statins with a focus on the liver [105]. They are the most effective medications for treating hypercholesterolemia and lowering cardiovascular disease-related mortality and morbidity. The microsomal 3-hydroxy-3-methylglutaryl-coenzyme a reductase inhibitor used to treat hypercholesterolemia is known as statins [106]. Additionally, a few recent investigations demonstrated extra-lipid effects and anticancer activities. Studies examining the anticancer effects of statins have recently increased due to the high prevalence of cancer disorders [107]. Creating a single, all-encompassing multistatin approach will be a convenient and cost-effective means of offering multiple statin analyses. Bridged Ethyl Hybrid (BEH) particle-based Acquity UPLC BEH C18 analytical column was used for chromatographic separation. ACN (A) and 0.5 mM ammonium acetate pH 4.0 (B) were used as the mobile phase in a gradient elution procedure with an initial ratio of 30:70 (A:B). On each of the three MS systems, the MS conditions were adjusted in both the positive and negative ESI modes. Because atorvastatin and rosuvastatin are the two most often used statins, it was determined from the study of Vlckova et al. that the approach was currently only employed for the examination of blood samples from patients who had taken these medications. Multistatin techniques are anticipated to be used for studying human and laboratory animal samples [102] (Table 5).

Conclusion

It is evident from the above-cited reports that UHPLC-MS/MS has played a tremendous role in identification, detection and quantification of a wide variety of analytes with great accuracy and sensitivity. Although, it offers several types of advantages and adopted wide applicability; still it is associated with certain problems like the use of toxic solvents, troubleshooting and complex data processing systems. Its adaptability can be improved by employing greener and less toxic solvents. Nonetheless, steady gradation of UHPLC-MS/MS techniques besides data-handling routines are still obligatory for data preprocessing, statistical analysis, biomarker recognition impurity, and degraded products.

References

[1] Fallick, G. J. Practical methods of high-speed liquid chromatography, *Advances in Chromatography*, CRC Press, 2021, pp. 61-97.

[2] Priyambada, P., G. Anju, *HPLC Columns and Their Role in Analytical Method Development: A Review*, (2018).

[3] Sah, P., M. P. Chasta, G. Sharma, K. Kishore, *High Performance Liquid Chromatography* (HPLC).

[4] Kondeti, R. R., K. S. Mulpuri, B. Meruga, Advancements in column chromatography: A review, *World Journal of Pharmaceutical Sciences* (2014) 1375-1383.

[5] Poole, S. K., C. F. Poole, Separation methods for estimating octanol–water partition coefficients, *Journal of Chromatography B* 797 (2003) 3-19.

[6] Patel, D. C., M. F. Wahab, D. W. Armstrong, Z. S. Breitbach, Advances in high-throughput and high-efficiency chiral liquid chromatographic separations, *Journal of Chromatography A* 1467 (2016) 2-18.

[7] Guillarme, D., J. Ruta, S. Rudaz, J. L. Veuthey, New trends in fast and high-resolution liquid chromatography: a critical comparison of existing approaches, *Analytical and bioanalytical chemistry* 397 (2010) 1069-1082.

[8] Fekete, S., I. Kohler, S. Rudaz, D. Guillarme, Importance of instrumentation for fast liquid chromatography in pharmaceutical analysis, *Journal of Pharmaceutical and Biomedical Analysis* 87 (2014) 105-119.

[9] Singh, J., A. Mehta, Rapid and sensitive detection of mycotoxins by advanced and emerging analytical methods: A review, *Food science & nutrition* 8 (2020) 2183-2204.

[10] Jakimska, A., A. Kot-Wasik, J. Namieśnik, The current state-of-the-art in the determination of pharmaceutical residues in environmental matrices using hyphenated techniques, *Critical Reviews in Analytical Chemistry* 44 (2014) 277-298.

[11] Chawla, G., C. Ranjan, Principle, instrumentation, and applications of UPLC: A novel technique of liquid chromatography, *Open Chemistry Journal* 3 (2016).

[12] Taleuzzaman, M., S. Ali, S. Gilani, S. Imam, A. Hafeez, Ultra performance liquid chromatography (UPLC)-a review, *Austin J Anal Pharm Chem* 2 (2015) 1056.

[13] Naresh, K., S. Bhawani, T. M. Kumar, *Ultra performance liquid chromatography,* System 1 (2014) 1.

[14] Sheliya, K., K. Shah, Ultra Performance Liquid Chromatography (UPLC): A Modern Chromatography Technique, *Pharma Science Monitor* 4 (2013).

[15] Nováková, L., L. Matysová, P. Solich, Advantages of application of UPLC in pharmaceutical analysis, *Talanta* 68 (2006) 908-918.

[16] Meenakshi, D., M. Meenaxi, A Review on Ultra Performance Liquid Chromatography, *International Journal of Drug Development and Research* 5 (2013) 0-0.

[17] Fountain, K. J., P. C. Iraneta, Instrumentation and columns for UHPLC separations, *UHPLC in Life Sciences* (2012) 29-66.

[18] Guillarme, D., J. L. Veuthey, Theory and Practice of UHPLC and UHPLC–MS, *Handbook of Advanced Chromatography/Mass Spectrometry Techniques*, Elsevier, 2017, pp. 1-38.

[19] Nováková, L., P. Svoboda, J. Pavlík, Ultra-high performance liquid chromatography, *Liquid chromatography*, Elsevier, 2017, pp. 719-769.

[20] Lauber, M. A., S. M. Koza, S. A. McCall, B. A. Alden, P. C. Iraneta, K. J. Fountain, High-resolution peptide mapping separations with MS-friendly mobile phases and charge-surface-modified C18, *Analytical chemistry* 85 (2013) 6936-6944.

[21] Fekete, S., A. Beck, E. Wagner, K. Vuignier, D. Guillarme, Adsorption and recovery issues of recombinant monoclonal antibodies in reversed-phase liquid chromatography, *Journal of Separation Science* 38 (2015) 1-8.

[22] Rathod, R. H., S. R. Chaudhari, A. S. Patil, A. A. Shirkhedkar, Ultra-high performance liquid chromatography-MS/MS (UHPLC-MS/MS) in practice: analysis of drugs and pharmaceutical formulations, *Future Journal of Pharmaceutical Sciences* 5 (2019) 1-26.

[23] Swartz, M. E. UPLC™: an introduction and review, *Journal of Liquid Chromatography & Related Technologies* 28 (2005) 1253-1263.

[24] da Silva, A. C. C., L. d. L. F. Lizot, M. F. Bastiani, M. V. Antunes, N. Brucker, R. Linden, Dried plasma spots for therapeutic monitoring of amikacin: Validation of an UHPLC-MS/MS assay and pharmacokinetic application, *Journal of Pharmaceutical and Biomedical Analysis* 184 (2020) 113201.

[25] Wang, Y., J. Ma, E. D. Martinez, D. Liang, H. Xie, A UHPLC-MS/MS method for the quantification of JIB-04 in rat plasma: Development, validation and application to pharmacokinetics study, *Journal of pharmaceutical and biomedical analysis* 191 (2020) 113587.

[26] Niu, C., W. Ye, X. Cui, J. Sun, S. Xiao, G. Chen, S. Bao, R. Chen, UHPLC-MS/MS method for the quantification of aloin-A in rat plasma and its application to a pharmacokinetic study, *Journal of Pharmaceutical and Biomedical Analysis* 178 (2020) 112928.

[27] Xie, L., X. Liu, X. Zhu, Y. Xu, S. Peng, K. Sun, H. Cai, Q. Dai, C. Wang, Q. Zhou, Development of an UHPLC-MS/MS method for comparative pharmacokinetics of nine anthraquinones in rats and application to dosage conversion between different Semen Cassiae forms, *Journal of Pharmaceutical and Biomedical Analysis* 174 (2019) 696-706.

[28] Chen, X., P. Zhu, B. Liu, L. Wei, Y. Xu, Simultaneous determination of fourteen compounds of *Hedyotis diffusa* Willd extract in rats by UHPLC–MS/MS method: Application to pharmacokinetics and tissue distribution study, *Journal of pharmaceutical and biomedical analysis* 159 (2018) 490-512.

[29] Eisenmann, E. D., Y. Jin, R. H. Weber, A. Sparreboom, S. D. Baker, Development and validation of a sensitive UHPLC-MS/MS analytical method for venetoclax in mouse plasma, and its application to pharmacokinetic studies, *Journal of Chromatography B* 1152 (2020) 122176.

[30] Parker, S. L., Y. C. Guerra Valero, D. M. Roberts, J. Lipman, J. A. Roberts, S. C. Wallis, Determination of cefalothin and cefazolin in human plasma, urine and peritoneal dialysate by UHPLC-MS/MS: Application to a pilot pharmacokinetic study in humans, *Biomedical Chromatography* 30 (2016) 872-879.

[31] Lindahl, S., R. Dyrkorn, O. Spigset, S. Hegstad, Quantification of apixaban, dabigatran, edoxaban, and rivaroxaban in human serum by UHPLC-MS/MS—Method development, validation, and application, *Therapeutic drug monitoring* 40 (2018) 369.

[32] Zhu, L. J., L. Chen, C. F. Bai, A. G. Wu, S. C. Liang, F. H. Huang, S. S. Cao, L. Yang, W. J. Zou, J. M. Wu, A rapid and sensitive UHPLC-MS/MS method for the determination of ziyuglycoside I and its application in a preliminary pharmacokinetic study in healthy and leukopenic rats, *Biomedicine & Pharmacotherapy* 123 (2020) 109756.

[33] Wang, P., J. Sun, E. Gao, Y. Zhao, W. Qu, Z. Yu, Simultaneous determination of limonin, dictamnine, obacunone and fraxinellone in rat plasma by a validated UHPLC–MS/MS and its application to a pharmacokinetic study after oral administration of Cortex Dictamni extract, *Journal of Chromatography B* 928 (2013) 44-51.

[34] Sun, Z., L. Zhao, L. Zuo, C. Qi, P. Zhao, X. Hou, A UHPLC–MS/MS method for simultaneous determination of six flavonoids, gallic acid and 5, 8-dihydroxy-1, 4-naphthoquinone in rat plasma and its application to a pharmacokinetic study of Cortex *Juglandis mandshuricae* extract, *Journal of Chromatography B* 958 (2014) 55-62.

[35] Zhao, L., Z. Xiong, Y. Sui, H. Zhu, Z. Zhou, Z. Wang, Y. Zhao, W. Xiao, J. Lin, K. Bi, Simultaneous determination of six bioactive constituents of Guizhi Fuling Capsule in rat plasma by UHPLC–MS/MS: Application to a pharmacokinetic study, *Journal of chromatography B* 1001 (2015) 49-57.

[36] Ji, B., Y. Zhao, Q. Zhang, P. Wang, J. Guan, R. Rong, Z. Yu, Simultaneous determination of cinnamaldehyde, cinnamic acid, and 2-methoxy cinnamic acid in rat whole blood after oral administration of volatile oil of Cinnamoni Ramulus by UHPLC-MS/MS: An application for a pharmacokinetic study, *Journal of Chromatography B* 1001 (2015) 107-113.

[37] Ramirez, M. S., M. E. Tolmasky, Amikacin: uses, resistance, and prospects for inhibition, *Molecules* 22 (2017) 2267.

[38] Grindlay, D., T. Reynolds, The Aloe vera phenomenon: a review of the properties and modern uses of the leaf parenchyma gel, *Journal of ethnopharmacology* 16 (1986) 117-151.

[39] Liu, F. W., F. C. Liu, Y. R. Wang, H. I. Tsai, H. P. Yu, Aloin protects skin fibroblasts from heat stress-induced oxidative stress damage by regulating the oxidative defense system, *PLoS One* 10 (2015) e0143528.

[40] Kumar, R., A. K. Singh, A. Gupta, A. Bishayee, A. K. Pandey, Therapeutic potential of *Aloe vera*—A miracle gift of nature, *Phytomedicine* 60 (2019) 152996.

[41] Dong, X., J. Fu, X. Yin, C. Yang, X. Zhang, W. Wang, X. Du, Q. Wang, J. Ni, Cassiae semen: A review of its phytochemistry and pharmacology, *Molecular Medicine Reports* 16 (2017) 2331-2346.

[42] Xie, W., Y. Zhao, L. Du, Emerging approaches of traditional Chinese medicine formulas for the treatment of hyperlipidemia, *Journal of Ethnopharmacology* 140 (2012) 345-367.

[43] Parveen, A., R. Sultana, S. M. Lee, T. H. Kim, S. Y. Kim, Phytochemicals against anti-diabetic complications: targeting the advanced glycation end product signaling pathway, *Archives of Pharmacal Research* 44 (2021) 378-401.

[44] Zhang, R., C. Ma, Y. Wei, X. Wang, J. Jia, J. Li, K. Li, G. Cao, P. Yang, Isolation, purification, structural characteristics, pharmacological activities, and combined action of *Hedyotis diffusa* polysaccharides: A review, *International journal of biological macromolecules* 183 (2021) 119-131.

[45] Zhu, H., A. Almasan, Development of venetoclax for therapy of lymphoid malignancies, *Drug design, development and therapy* 11 (2017) 685.

[46] Duff, P. Prophylactic antibiotics for cesarean delivery: a simple cost-effective strategy for prevention of postoperative morbidity, *American journal of obstetrics and gynecology* 157 (1987) 794-798.
[47] Dalrymple, L. S., A. S. Go, Epidemiology of acute infections among patients with chronic kidney disease, *Clinical Journal of the American Society of Nephrology* 3 (2008) 1487-1493.
[48] Vimalesvaran, K., S. J. Dockrill, D. A. Gorog, Role of rivaroxaban in the management of atrial fibrillation: insights from clinical practice, *Vascular Health and Risk Management* 14 (2018) 13.
[49] Adcock, D., R. Gosselin, Direct oral anticoagulants (DOACs) in the laboratory: 2015 review, *Thrombosis research* 136 (2015) 7-12.
[50] Wang, L., H. Li, X. Shen, J. Zeng, L. Yue, J. Lin, J. Yang, W. Zou, Y. Li, D. Qin, Elucidation of the molecular mechanism of *Sanguisorba officinalis* L. against leukopenia based on network pharmacology, *Biomedicine & Pharmacotherapy* 132 (2020) 110934.
[51] Xiong, Y., Y. Zou, L. Chen, Y. Xu, S. Wang, Development and *in vivo* evaluation of ziyuglycoside i–loaded self-microemulsifying formulation for activity of increasing leukocyte, *AAPS PharmSciTech* 20 (2019) 1-10.
[52] Zhao, P., C. Qi, G. Wang, X. Dai, X. Hou, Enrichment and purification of total flavonoids from Cortex *Juglandis mandshuricae* extracts and their suppressive effect on carbon tetrachloride-induced hepatic injury in Mice, *Journal of Chromatography B* 1007 (2015) 8-17.
[53] Zhao, P., L. Zhao, C. Qi, G. Wang, X. Hou, A hierarchical cluster analysis of ten index constituents based on microwave-assisted extraction by UHPLC-MS/MS for the evaluation and quality control of Cortex *Juglandis mandshuricae*, *Analytical Methods* 7 (2015) 1816-1824.
[54] Liu, S., L. Yang, S. Zheng, A. Hou, W. Man, J. Zhang, S. Wang, X. Wang, H. Yu, H. Jiang, A review: the botany, ethnopharmacology, phytochemistry, pharmacology of Cinnamomi cortex, *RSC advances* 11 (2021) 27461-27497.
[55] Qin, N., M. Shen, P. Xiang, D. Wen, B. Shen, H. Deng, H. Qiang, F. Song, Y. Shi, Determination of 37 fentanyl analogues and novel synthetic opioids in hair by UHPLC-MS/MS and its application to authentic cases, *Scientific Reports* 10 (2020) 1-13.
[56] Wang, X., S. S. Johansen, M. K. K. Nielsen, K. Linnet, Targeted analysis of 116 drugs in hair by UHPLC-MS/MS and its application to forensic cases, *Drug testing and analysis* 9 (2017) 1137-1151.
[57] Odoardi, S., L. Anzillotti, S. Strano-Rossi, Simplifying sample pretreatment: Application of dried blood spot (DBS) method to blood samples, including postmortem, for UHPLC–MS/MS analysis of drugs of abuse, *Forensic science international* 243 (2014) 61-67.
[58] Shi, Y., R. Wang, S. Yuan, H. Qiang, M. Shen, B. Shen, O. H. Drummer, Z. Yu, Y. Zhao, P. Xiang, UHPLC-MS/MS method for simultaneously detecting 16 tryptamines and their metabolites in human hair and applications to real forensics cases, *Journal of Chromatography B* 1159 (2020) 122392.

[59] Busardò, F. P., C. Kyriakou, E. Marchei, R. Pacifici, D. S. Pedersen, S. Pichini, Ultra-high performance liquid chromatography tandem mass spectrometry (UHPLC–MS/MS) for determination of GHB, precursors and metabolites in different specimens: application to clinical and forensic cases, *Journal of pharmaceutical and biomedical analysis* 137 (2017) 123-131.

[60] Anzillotti, L., E. Castrignanò, S. S. Rossi, M. Chiarotti, Cannabinoids determination in oral fluid by SPME–GC/MS and UHPLC–MS/MS and its application on suspected drivers, *Science & Justice* 54 (2014) 421-426.

[61] Mannocchi, G., A. Di Trana, A. Tini, S. Zaami, M. Gottardi, S. Pichini, F. P. Busardò, Development and validation of fast UHPLC-MS/MS screening method for 87 NPS and 32 other drugs of abuse in hair and nails: application to real cases, *Analytical and Bioanalytical Chemistry* 412 (2020) 5125-5145.

[62] Lovrecic, B., M. Lovrecic, B. Gabrovec, M. Carli, M. Pacini, A. G. Maremmani, I. Maremmani, Non-medical use of novel synthetic opioids: a new challenge to public health, *International journal of environmental research and public health* 16 (2019) 177.

[63] Withers, C. E., M. S. Scheller, *Anesthesia for interventional radiologic procedures, Seminars in interventional radiology,* Copyright© 1988 by Thieme Medical Publishers, Inc., 1988, pp. 125-131.

[64] Guthrie, R., A. Susi, A simple phenylalanine method for detecting phenylketonuria in large populations of newborn infants, *Pediatrics* 32 (1963) 338-343.

[65] Stanley, T. H., T. D. Egan, H. Van Aken, A tribute to Dr. Paul AJ Janssen: entrepreneur extraordinaire, innovative scientist, and significant contributor to anesthesiology, *Anesthesia & Analgesi*a 106 (2008) 451-462.

[66] Araújo, A. M., F. Carvalho, M. d. L. Bastos, P. Guedes de Pinho, M. Carvalho, The hallucinogenic world of tryptamines: an updated review, *Archives of toxicology* 89 (2015) 1151-1173.

[67] Tittarelli, R., G. Mannocchi, F. Pantano, F. Saverio Romolo, Recreational use, analysis and toxicity of tryptamines, *Current neuropharmacology* 13 (2015) 26-46.

[68] Snead III, O. C., K. M. Gibson, γ-Hydroxybutyric acid, *New England Journal of Medicine* 352 (2005) 2721-2732.

[69] Tay, E., W. K. W. Lo, B. Murnion, *Current Insights on the Impact of Gamma-Hydroxybutyrate (GHB) Abuse, Substance Abuse and Rehabilitation* 13 (2022) 13.

[70] Pichini, S., S. Malaca, M. Gottardi, A. P. Pérez-Acevedo, E. Papaseit, C. Perez-Maña, M. Farré, R. Pacifici, A. Tagliabracci, G. Mannocchi, UHPLC-MS/MS analysis of cannabidiol metabolites in serum and urine samples. Application to an individual treated with medical cannabis, *Talanta* 223 (2021) 121772.

[71] Zhao, M., T. Zhang, G. Li, F. Qiu, Y. Sun, L. Zhao, Simultaneous determination of valproic acid and its major metabolites by UHPLC-MS/MS in Chinese patients: Application to therapeutic drug monitoring, *Journal of chromatographic science* 55 (2017) 436-444.

[72] Yu, C., R. Liu, C. Xie, Q. Zhang, Y. Yin, K. Bi, Q. Li, Quantification of free polyamines and their metabolites in biofluids and liver tissue by UHPLC-MS/MS: application to identify the potential biomarkers of hepatocellular carcinoma, *Analytical and bioanalytical chemistry* 407 (2015) 6891-6897.

[73] Goessens, T., S. De Baere, N. De Troyer, A. Deknock, P. Goethals, L. Lens, F. Pasmans, S. Croubels, Multi-residue analysis of 20 mycotoxins including major metabolites and emerging mycotoxins in freshwater using UHPLC-MS/MS and application to freshwater ponds in flanders, Belgium, *Environmental Research* 196 (2021) 110366.

[74] Li, J. P., J. M. Guo, E. X. Shang, Z. H. Zhu, Y. Liu, B. C. Zhao, J. Zhao, Z. S. Tang, J. A. Duan, Quantitative determination of five metabolites of aspirin by UHPLC–MS/MS coupled with enzymatic reaction and its application to evaluate the effects of aspirin dosage on the metabolic profile, *Journal of Pharmaceutical and Biomedical Analysis* 138 (2017) 109-117.

[75] Schoeman, J. C., A. C. Harms, M. van Weeghel, R. Berger, R. J. Vreeken, T. Hankemeier, Development and application of a UHPLC–MS/MS metabolomics based comprehensive systemic and tissue-specific screening method for inflammatory, oxidative and nitrosative stress, *Analytical and bioanalytical chemistry* 410 (2018) 2551-2568.

[76] Cavalheiro, E. K. F. F., A. B. Costa, D. H. Salla, M. R. d. Silva, T. F. Mendes, L.E.d. Silva, C. d. R. Turatti, R. M. d. Bitencourt, G. T. Rezin, Cannabis sativa as a treatment for obesity: From anti-inflammatory indirect support to a promising metabolic re-establishment target, *Cannabis and Cannabinoid Research* 7 (2022) 135-151.

[77] Scuderi, C., D. D. Filippis, T. Iuvone, A. Blasio, A. Steardo, G. Esposito, Cannabidiol in medicine: a review of its therapeutic potential in CNS disorders, *Phytotherapy Research: An International Journal Devoted to Pharmacological and Toxicological Evaluation of Natural Product Derivatives* 23 (2009) 597-602.

[78] Wlodarczyk, B. J., A. M. Palacios, T. M. George, R. H. Finnell, Antiepileptic drugs and pregnancy outcomes, *American Journal of Medical Genetics Part A* 158 (2012) 2071-2090.

[79] Acid, V. *Valproic Acid*, (2003).

[80] Alshannaq, A., J. H. Yu, Occurrence, toxicity, and analysis of major mycotoxins in food, *International journal of environmental research and public health* 14 (2017) 632.

[81] Haruna, M., S. Bashir, M. Hassan, S. Ibrahim, I. Safiyanu, A. Tijjani, *Comparative Proximate Analysis of Fried and Roasted Tilapia (Oreochromis niloticus) Fish.*

[82] Yeomans, N. D., C. J. Hawkey, W. Brailsford, J. Næsdal, Gastroduodenal toxicity of low-dose acetylsalicylic acid: a comparison with non-steroidal anti-inflammatory drugs, *Current medical research and opinion* 25 (2009) 2785-2793.

[83] Zheng, Z., X. E. Zhao, S. Zhu, J. Dang, X. Qiao, Z. Qiu, Y. Tao, Simultaneous determination of oleanolic acid and ursolic acid by *in vivo* microdialysis via UHPLC-MS/MS using magnetic dispersive solid phase extraction coupling with microwave-assisted derivatization and its application to a pharmacokinetic study of *Arctiumlappa* L. root extract in rats, *Journal of agricultural and food chemistry* 66 (2018) 3975-3982.

[84] Lachat, L., G. T. Glauser, Development and validation of an ultra-sensitive UHPLC–MS/MS method for neonicotinoid analysis in milk, *Journal of agricultural and food chemistry* 66 (2018) 8639-8646.

[85] Tian, Y., Q. Li, X. Zhou, Q. Pang, Y. Xu, A UHPLC–MS/MS method for simultaneous determination of twelve constituents from *Erigeron breviscapus* extract in rat plasma: application to a pharmacokinetic study, *Journal of Chromatography B* 1046 (2017) 1-12.

[86] Moradi-Afrapoli, F., M. Oufir, F. R. Walter, M. A. Deli, M. Smiesko, V. Zabela, V. Butterweck, M. Hamburger, Validation of UHPLC–MS/MS methods for the determination of kaempferol and its metabolite 4-hydroxyphenyl acetic acid, and application to *in vitro* blood-brain barrier and intestinal drug permeability studies, *Journal of pharmaceutical and biomedical analysis* 128 (2016) 264-274.

[87] Wang, Z., W. Sun, Z. F. Lin, R. Sun, C. K. Huang, W. J. Ye, Y. Y. Dong, X. D. Zhang, R. J. Chen, A UHPLC-MS/MS method coupled with liquid-liquid extraction for the quantitation of phenacetin, omeprazole, metoprolol, midazolam and their metabolites in rat plasma and its application to the study of four CYP450 activities, *Journal of Pharmaceutical and Biomedical Analysis* 163 (2019) 204-210.

[88] Yu, W., D. Wen, D. Cai, J. Zheng, H. Gan, F. Jiang, X. Liu, B. Lao, W. Yu, Y. Guan, Simultaneous determination of curcumin, tetrahydrocurcumin, quercetin, and paeoniflorin by UHPLC-MS/MS in rat plasma and its application to a pharmacokinetic study, *Journal of Pharmaceutical and Biomedical Analysis* 172 (2019) 58-66.

[89] Prokopienko, A. J., R. E. West III, J. R. Stubbs, T. D. Nolin, Development and validation of a UHPLC-MS/MS method for measurement of a gut-derived uremic toxin panel in human serum: an application in patients with kidney disease, *Journal of pharmaceutical and biomedical analysis* 174 (2019) 618-624.

[90] Nováková, L., A. Vildová, J. P. Mateus, T. Gonçalves, P. Solich, Development and application of UHPLC–MS/MS method for the determination of phenolic compounds in Chamomile flowers and Chamomile tea extracts, *Talanta* 82 (2010) 1271-1280.

[91] Genangeli, M., G. Caprioli, M. Cortese, F. Laus, M. Matteucci, R. Petrelli, M. Ricciutelli, G. Sagratini, S. Sartori, S. Vittori, Development and application of a UHPLC-MS/MS method for the simultaneous determination of 17 steroidal hormones in equine serum, *Journal of Mass Spectrometry* 52 (2017) 22-29.

[92] Yuan, Y., X. Qiu, D. Nikolic, J. H. Dahl, R. B. van Breemen, Method development and validation for ultra-high-pressure LC/MS/MS determination of hop prenylflavonoids in human serum, *Journal of AOAC International* 95 (2012) 1744-1749.

[93] Cambien, G., N. Venisse, V. Migeot, S. Rabouan, M. Belmouaz, G. Binson, M. Albouy-Llaty, S. Ayraud-Thevenot, A. Dupuis, Simultaneous determination of bisphenol A and its chlorinated derivatives in human plasma: Development, validation and application of a UHPLC–MS/MS method, *Chemosphere* 242 (2020) 125236.

[94] Shao, X., Z. Liu, X. Xu, Z. Li, X. Qian, Overall status of neonicotinoid insecticides in China: production, application and innovation, *Journal of Pesticide Science* (2012) D12-037.

[95] Goulson, D. An overview of the environmental risks posed by neonicotinoid insecticides, *Journal of Applied Ecology* 50 (2013) 977-987.

[96] Vassiliou, D., E. Sardh, P. Harper, A. R. Simon, V. A. Clausen, N. Najafian, G. J. Robbie, S. Agarwal, A Drug-Drug Interaction Study Evaluating the Effect of Givosiran, a Small Interfering Ribonucleic Acid, on Cytochrome P450 Activity in the Liver, *Clinical Pharmacology & Therapeutics* 110 (2021) 1250-1260.

[97] Tadbir, A. A., S. Pourshahidi, H. Ebrahimi, Z. Hajipour, M. R. Memarzade, S. Shirazian, The effect of *Matricaria chamomilla* (chamomile) extract in Orabase on minor aphthous stomatitis, a randomized clinical trial, *Journal of Herbal Medicine* 5 (2015) 71-76.

[98] Howdeshell, K. L., P. H. Peterman, B. M. Judy, J. A. Taylor, C. E. Orazio, R. L. Ruhlen, F. S. Vom Saal, W. V. Welshons, Bisphenol A is released from used polycarbonate animal cages into water at room temperature, *Environmental Health Perspectives* 111 (2003) 1180-1187.

[99] Kokoszka, K., J. Wilk, E. Felis, S. Bajkacz, Application of UHPLC-MS/MS method to study occurrence and fate of sulfonamide antibiotics and their transformation products in surface water in highly urbanized areas, *Chemosphere* 283 (2021) 131189.

[100] Gorissen, B., T. Reyns, M. Devreese, P. De Backer, J. Van Loco, S. Croubels, Determination of selected veterinary antimicrobials in poultry excreta by UHPLC-MS/MS, for application in Salmonella control programs, *Analytical and Bioanalytical Chemistry* 407 (2015) 4447-4457.

[101] Kehler, J. R., C. L. Bowen, S. L. Boram, C. A. Evans, Application of DBS for quantitative assessment of the peptide Exendin-4; comparison of plasma and DBS method by UHPLC–MS/MS, *Bioanalysis* 2 (2010) 1461-1468.

[102] Vlčková, H., P. Svoboda, O. Novák, P. Solich, L. Nováková, Development of MEPS-UHPLC–MS/MS multistatin methods for clinical analysis, *Bioanalysis 8* (2016) 333-349.

[103] Jafari Ozumchelouei, E., A. H. Hamidian, Y. Zhang, M. Yang, Physicochemical properties of antibiotics: A review with an emphasis on detection in the aquatic environment, *Water environment research* 92 (2020) 177-188.

[104] Agyare, C., V. E. Boamah, C. N. Zumbi, F. B. Osei, Antibiotic use in poultry production and its effects on bacterial resistance, *Antimicrobial resistance—A global threat* (2018) 33-51.

[105] Shitara, Y., Y. Sugiyama, Pharmacokinetic and pharmacodynamic alterations of 3-hydroxy-3-methylglutaryl coenzyme A (HMG-CoA) reductase inhibitors: drug–drug interactions and interindividual differences in transporter and metabolic enzyme functions, *Pharmacology & therapeutics* 112 (2006) 71-105.

[106] Christians, U., W. Jacobsen, L. C. Floren, Metabolism and drug interactions of 3-hydroxy-3-methylglutaryl coenzyme A reductase inhibitors in transplant patients: are the statins mechanistically similar? *Pharmacology & therapeutics* 80 (1998) 1-34.

[107] Sleijfer, S., A. van der Gaast, S. T. André, G. Stoter, J. Verweij, The potential of statins as part of anti-cancer treatment, *European journal of cancer* 41 (2005) 516-522.

Chapter 3

The Recent Advancements and Applications of Hydrophilic Interaction Liquid Chromatography (HILIC)

Yogindra Kumari
Arshdeep Chopra
and Rohit Bhatia*
Department of Pharmaceutical Analysis, ISF College of Pharmacy, Ghal Kalan, Moga, Punjab, India

Abstract

HILIC has recently gained popularity due to its ability to separate polar compounds. HILIC involves the separation of polar analytes through their partitioning into a water-rich layer formed on the surface of the stationary phase. This chapter illustrates the mode of operation of HILIC, a comparison of HILIC concerning normal phase liquid chromatography (NPLC) and reverse phase liquid chromatography (RPLC), instrumentation, and various applications of HILIC in the field of food, drug, agricultural, and adulterant analysis. Quantifying drugs and related compounds in complex matrices, including biological and environmental samples, pharmaceutical formulations, food, and animal feed, provide higher selectivity, sensitivity, and efficiency. HILIC is suitable for analyzing substances in complex processes which elute close to the void in reverse-phase chromatography. HILIC does not require ion-pair reagents, which makes it convenient to couple with MS. Endogenous metabolite quantification is a growing field in drug discovery, with

* Corresponding Author's Email: bhatiarohit5678@gmail.com.

In: Modern Approaches in Fluid Chromatography
Editors: Rohit Bhatia and Bhupinder Kumar
ISBN: 979-8-88697-856-8

applications ranging from assessing pharmacokinetic-pharmacodynamic relationships to targeting activity and efficacy.

Keywords: hydrophilic interaction liquid chromatography, normal phase, instrumentation, ion-pair

1. Introduction

New analytical methods that can handle many analytes in complex matrices are urgently needed. Hydrophilic Interaction Liquid Chromatography (HILIC) has recently been demonstrated as an essential supplement to reversed-phase liquid chromatography for polar analytes, particularly endogenous compounds [1]. A.J. Alpert introduced the term HILIC in 1990, a variant of Normal Phase Liquid Chromatography (NPLC). HILIC can separate polar and hydrophilic solutes with a polar material and an aqueous-organic mobile phase, generally called Aqueous HILIC, for which acetonitrile is mainly used [2-4]. The solvent systems used for HILIC are the same as used in reversed-phase (RP) separations, i.e., chloroform, cyclohexane, benzene, carbon tetrachloride, etc. Adopting a wide variety of HILIC columns from commercial or research sources has facilitated the fast adoption of this technique [5]. HILIC also allows the analysis of charged substances, as in Ion Chromatography (IC). Figure 1 shows how HILIC complements other chromatographic techniques and extends the range of separation options [6]. Examples of stationary phases commonly used in HILIC are unmodified silica or silica phases modified with amino, amide, siloxanes, silanols, zwitterionic functional groups, polyols including saccharides, and other polar groups. HILIC is extensively applied in various fields, including pharmaceutical chemistry, proteomics, metabolomics, food and agricultural chemistry, medicinal chemistry, and many others [7].

Hyphenation of RPLC and HILIC is also employed to achieve higher separation ability. Even though acetonitrile (ACN) and water are used as the same solvents in both RPLC and HILIC, it can be challenging to perform hyphenation between these two methods since HILIC uses a more potent elution solvent than RPLC and vice-versa. To increase metabolite coverage in contrast to a single or HILIC column, serially connected RPLC and HILIC have been developed by positioning dilution pump(s) at the RP column's outlet. However, there hasn't been a significant increase in peak capacity and the signals formed gather at relatively narrow regions of the chromatogram

[8]. Figure 1 depicts the integration of HILIC with other types of chromatography.

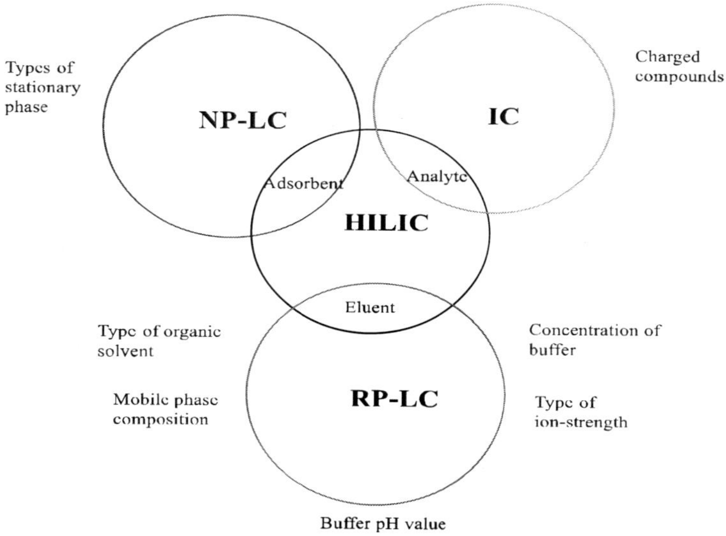

Figure 1. HILIC combines the characteristics of the three methods in liquid chromatography.

1.1. Mode of Operation

There is an ongoing discussion on the HILIC mode's mechanism. Alpert stated that the chief separation instrument involved is partitioning the analyte between the aqueous layer on the polar stationary and mobile phases (Figure 2). Specifically, water molecules solvate the polar groups bound to the stationary phase, generating an aqueous layer on the surface. The semi-immobilized aqueous layer and the primary organic mobile phase, which has some aqueous content, partition the analyte, dissolved in the mobile phase, between them. A polar solute has a higher affinity for the semi-immobilized aqueous layer than the leading organic mobile phase, as it is better solvated in the former. This preference causes the solute to interact with the aqueous layer more frequently, boosting retention [9].

However, it appears that the partitioning mechanism is not the only one involved in the HILIC separations. Another mechanism involved is the adsorption mechanism (i.e., dipole-dipole interactions, hydrogen bonding interaction of neutral polar species, and electrostatic interactions of ionized species - attractive and repulsive). The aqueous layer is dynamically immobilized over the hydrophilic stationary phases. Small-sized analytes could easily penetrate this aqueous layer during the equilibrium to interact with the surface functional groups, such as silanols, amino, or zwitterionic groups, showing that an adsorption process played a role in HILIC retention [9, 10]. Under distinct stationary phases and chromatographic circumstances, various analytes exhibit various retention behaviors, the actual retention mechanism depends on the mobile phase composition, the analyte characteristics (polarity and charge status), and the functional groups of the stationary phases.

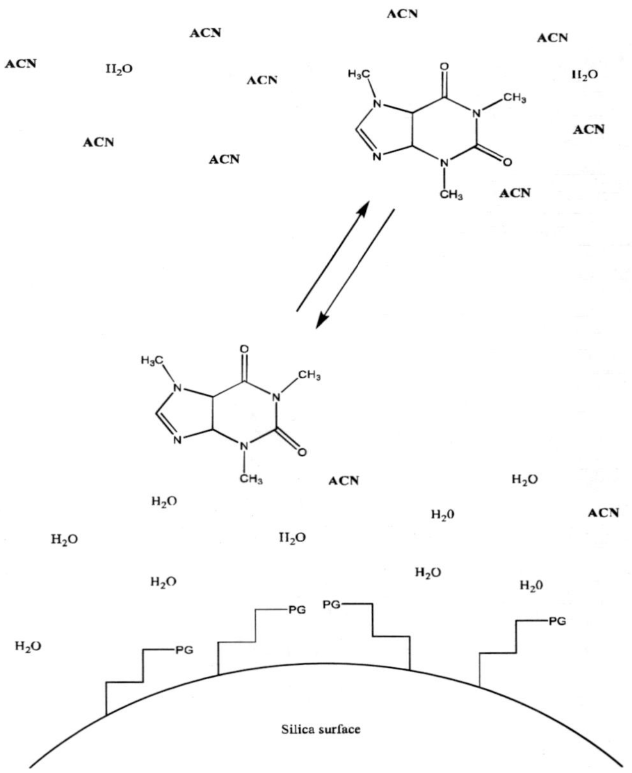

Figure 2. The water-enriched layer formed over the polar stationary phase, showing interactions between the analytes and the two liquid phases. PG is the Polar Group attached to the stationary phase.

1.2. HILIC in Contrast to RPLC and NPLC

Since the advent of the bonded phases, reversed-phase liquid chromatography (RPLC) has grown to be one of the most effective analytical procedures due to its potential for separation, versatility and reliability. However, due to their poor retention, the analysis of highly polar compounds by RPLC remains challenging [11]. HILIC exhibits a complementary selectivity to RPLC, providing an elution order opposite to that obtained in RPLC, and in most cases, proves to be the best option for separating compounds that cannot be sufficiently retained under RPLC conditions. Another advantage is using water-miscible organic-rich mobile phases, which are generally a combination of water and acetonitrile. Another positive aspect of using these liquid phases is the lower back pressure than RPLC mobile phases due to the minimal viscosity of the organic-rich eluents and the ease of injecting samples in organic solvents [12]. To successfully separate polar substances, the conventional C_{18} bonded phases can be replaced with more polar phases where water is adsorbed in HILIC.

Moreover, HILIC has established itself as the separation method of choice for uncharged highly hydrophilic and amphiphilic chemicals that are too polar to be well separated in RPLC and may not have enough charge to allow for efficient electrostatic retention in IC [6]. The retention mechanism in HILIC is more complex when compared to RPLC, simultaneously comprising several kinds of interactions. In most instances, the solutes least retained in the reversed-phase are the most contained in HILIC [13].

In NPLC, the elution order is similar, but the polar compounds always have good solubility in the hydro-organic mobile phase employed in HILIC than in the organic mobile phase of NPLC, which outweighs the disadvantage of the frequently observed poor solubility in NPLC [13]. In NPLC, hexane-based non-aqueous apolar mobile phases are used, and the addition of polar solvents such as ethyl acetate facilitates the elution. HILIC mobile phases, in contrast to NPLC, are blends of organic solvents that are water-water miscible, typically acetonitrile [12]. This property not only helps to solve the issue of low aqueous solubility but also increases the sensitivity and detectability of HILIC for MS [14]. The fact that water is the most potent eluting solvent makes it another factor that sets HILIC apart from other separation mechanisms, such as NPLC and means that as the water content in the mobile phase increases, the retention of analytes falls [15].

1.3. Recent Advancements in HILIC

The need for chromatographic techniques compatible with Mass Spectrometry (MS) that can separate intact proteins has increased and recent studies have indicated that Hydrophilic Interaction Liquid Chromatography (HILIC) holds a good potential for characterizing glycoforms of intact proteins [16]. Utilizing HILIC stationary phases enables higher concentrations of organic solvent in the mobile phase, which is favorable for mass spectrometric ionization, especially ESI [15]. Various column chemistries (and hence selectivities) have been applied to separate proteins and proteoforms. HILIC-MS has been successfully employed to study proteins such as membrane proteins, histones, intact and IdeS-digested monoclonal antibodies, biopharmaceuticals, and neoglycoproteins. However, the use of HILIC-MS for the analysis of low-abundant proteins is currently constrained by two main problems: the suppression of protein ionization by mobile-phase additives necessary for protein elution and challenges with sample and eluent solvent compatibility [16]. The most often used detection methods are mass spectrometric and diode array (DAD) detection [15].

The separation and identification of phosphorothioate oligonucleotides using a combination of Hydrophilic Interaction Liquid Chromatography (HILIC) and Electrospray Ionisation-Mass Spectrometry (ESI-MS) method have also been carried out. Using HILIC, phosphorothioate 21-mer and 23-mer were separated, and Selected Ion Monitoring (SIM) ESI-MS was used to detect them [17].

Procedures utilizing SPE and HILIC with tandem MS (HILIC-MS/MS) are frequently used for detecting and quantifying pollutants in environmental matrices. To assess the most significant (Gadolinium) Gd-based MRI contrast agents, HILIC and Inductively Coupled Plasma Mass Spectrometry (HILIC-ICP/MS) have been combined and applied in hospital effluents and Waste Water Treatment Plant (WWTP) sewage [7].

HILIC has been employed with either untargeted MS detection or with MS/MS for targeted profiling of biological samples (urine, blood) or samples from plant and food origin, but still, there are limitations in the analysis of polar metabolites. Using HILIC or ion-pair LC for separation and triple quadrupole or Quadrupole-Time of Flight (Q-ToF) MS technology for analyte quantification, many groups have focused on studying extracellular culture medium and intracellular metabolites [18].

As demonstrated by Gilar *et al.*, the orthogonality of separation selectivity in 2D-LC coupling HILIC and RPLC was higher than in any other combination investigated [19]. It was concluded that 2D-LC of HILIC-RPLC would make an effective separation method for complex analytes, such as those used in metabolomics analysis. N-glycans, phospholipids, and peptides were all analyzed using 2D-LC. Under HILIC conditions, 1-D separation of phospholipids into their respective classes was completed. Then the initially acquired fractions were further separated and identified in the RPLC-MS system in the second dimension. Hence, HILIC-RPLC-MS is an excellent combination for complex analytes [1].

Liquid Chromatography (LC) and Nuclear Magnetic Resonance (NMR) have demonstrated significant value in several applications. HILIC has been combined with isotope-tagged NMR to identify carboxyl-containing metabolites in human urine. The use of HILIC-NMR for the separation and identification of chemically derivatized metabolites in human urine incorporates the advantages of conventional LC-NMR of bio-fluids (such as reduced spectra complexity) while minimizing the effects of pH-dependent chemical shift variations and the consequent solvent masking of lower intensity signals [20].

An at-line automatic micro-Solid Phase Extraction (μSPE) method for the determination of salivary cotinine (which is the first metabolite of nicotine) is often used as a biomarker in monitoring environmental tobacco smoke exposure) followed by its analysis using HILIC has been used. Due to the relatively high polarity of the cotinine molecule, HILIC was used to increase the sensitivity of the recently developed micro-Solid Phase Extraction-Bead Injection-Lab-on-Value (μSPE-BI-LOV technique) [21].

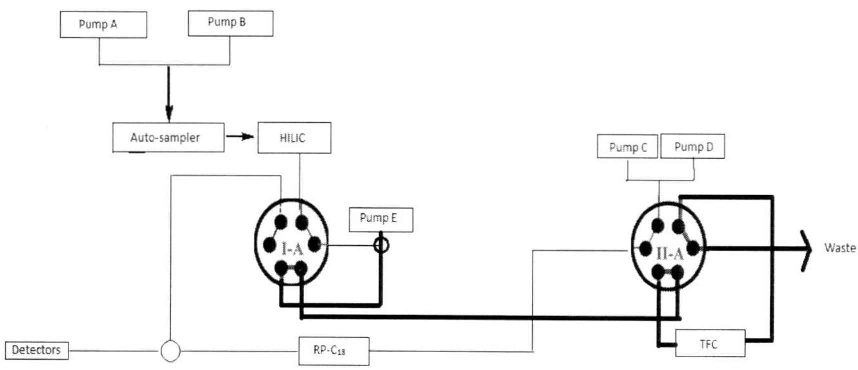

Figure 3. Instrumentation of HILIC technique.

1.4. Instrumentation of HILIC

Figure 3 depicts the typical instrumentation involved in HILIC technology.

1.4.1. Stationary Phases Used in HILIC

Any polar chromatographic surface may be utilized for HILIC separations. Classical bare silica or silica gels that have been heavily modified with polar functional groups make up most of HILIC stationary phases. Polymer-based stationary phases can also be employed. The first generation of HILIC mode separations began in 1975. In an acetonitrile and water (75:25 v/v) mixture, Linden *et al.* isolated carbohydrates using an amino-silica phase, Bondapak [22]. The next generation of stationary phases for HILIC utilized amide- and DIOL-silica. Protein separation has been the primary use for the DIOL-silica column [23]. HILIC has advanced into second and third-generation systems during the past 15 years, most incorporating mixed or multiple-interaction solid phases [24]. Based on the charge properties of the functional groups, HILIC stationary phases are divided into neutral, charged, and zwitterionic phases.

Neutral Stationary Phases

The functional groups in this category—such as amide, cyano, diol, and cyclodextrin—cannot be charged at the pH range usual for the mobile phase in HILIC. The amide phase is one of the most well-liked phrases in this category and has been used extensively in HILIC [25-28]. The amide molecule is joined to the silica surface using a propyl linker or a unique linkage. The aspartamide phase deserves special attention because it was the first stationary phase created specifically for HILIC separations. A layer of polysuccinimide is fused to an aminopropylated silica layer to create the aspartamide phase, which is subsequently treated with ethanolamine to make the stationary phase. The aspartamide phase has separated proteins, peptides, and small polar molecules [29-31]. However, it is not utilized as frequently as the amide phase, maybe due to its reduced efficiency and shorter long-lived stability. This group's cyano and diol phases are commonly employed in normal phase separations, although their applications in HILIC are still somewhat confined. Cyano groups are less hydrophilic and cannot act as hydrogen bond donors. Due to this, most polar chemicals cannot be retained effectively, and only a few unique applications for the cyano phase have been found [32].

In contrast, the diol phase has more excellent HILIC applications, maybe due to more hydrophilic dihydroxyl groups and the ability to form hydrogen bonds [33-36]. In chiral stationary phases for enantiomeric separations, cyclodextrins (CDs) are frequently utilized [37]. The rims of CDs' truncated cone structure include hydroxyl groups, which enable the use of the CD phase for HILIC separations [38]. Polar compounds and oligosaccharides have been separated using the neutral CD phases. Additionally, it has been demonstrated that the CD phase can resolve polar chiral molecules in the HILIC mode [39, 40].

Charged Stationary Phases

The functional groups of the stationary phases in this category can carry charges in the normal HILIC pH range (e.g., pH 3-7). A typical example in this category is the amino phase, which has been employed extensively for HILIC separations [24, 41, 42]. After end-capping, an innovative amino phase called TSKgel NH2-100 (Tosoh Bioscience) is prepared. There have also been limited HILIC separations using polyamine phases such as YMC Polyamine II, PolyWAX LP, and Luna NH2 [43-45]. Imidazole and triazole phases are two more recent additions to the charged phase group. Both the monomeric phases are prepared on silica packing materials and might be positively charged in the typical pH range for separation in HILIC [46]. A few negatively charged phases have also been utilized for HILIC separation in addition to positively charged phases, like poly (2-sulfoethylaspartamide), poly (aspartic acid) and sulfonated S-DVB phase. Under HILIC conditions, these phases have been used to separate peptides and essential medicinal molecules, generally utilized as cation exchangers in ion-exchange chromatography. In HILIC, positive-charged species (e.g., amino and polyamine phases), are used in more numbers than negatively-charged species [27].

Zwitterionic Phases

A relatively new class of polar stationary phases utilized for HILIC separation are zwitterionic phases (for ex- sulfobetaine phase) that developed initially for separating inorganic salts and more minor ionic compounds but still managed to find a wide range of applications in HILIC [24, 41, 47]. The negative charge of the sulfonic group at the distal end of the sulfobetaine phase might cause electrostatic interactions with the direction solutes [45, 48]. The phosphocholine base is based on a graft polymer with zwitterionic phosphorylcholine groups attached to its surface. The main distinction between the two zwitterionic phases is that the positively charged quaternary ammonium group is at the distal end of the phosphocholine phase, whereas the negatively charged sulfonate group is at the distal end of the sulfobetaine phase [27, 49].

1.4.2. Mobile Phases Used in Hilic

Organic-rich mobile phases, typically containing 5–40% water or a buffer, are used in HILIC on polar stationary phases. To ensure the elution of all sample components, it is advised to conduct early studies at a relatively high concentration (for example, 40%) of aqueous buffer in acetonitrile. The elution intensity is then controlled by boosting the acetonitrile concentration until suitable sample retention is attained. Water is firmly bound to the surface of charged or uncharged polar adsorbents from the mixed aqueous-organic mobile phase. In mobile phases with water concentrations more significant than a few hundred ppm, the preferred adsorption leads to forming a water-rich adsorbed diffuse liquid multi-layer [50]. The selection of the organic solvent significantly affects the retention in the HILIC mode. The order of increasing solvent polarity and ability to engage in proton-donor/proton-acceptor interactions affects the elution strength of organic solvents in the HILIC mode i.e., the elution strength increases with increasing solvent polarity: methanol > ethanol > 2-propanol > tetrahydrofuran > acetonitrile [51].

In most HILIC applications, acetonitrile is the organic solvent of choice since mobile phases containing other solvents frequently result in poor sample retention and asymmetric or broad peak shapes. This applies particularly to methanol, which is rarely employed as the organic solvent in HILIC applications. Several reported attempts have been made to replace acetonitrile with a less toxic solvent, but this effort has not been very successful, except for a few instances where acetonitrile can be substituted with tetrahydrofuran [52].

The retention and selectivity of nucleic bases, cortisol, flurbiprofen, theophylline, and caffeine in HILIC on bare silica are affected by adding carbon dioxide to the acetonitrile-water and ethanol-water mobile phases. High carbon dioxide levels in ethanol were reported to produce separations similar to those attained with acetonitrile-water mobile phases [53]. Because HILIC columns are more likely to become overloaded than RP columns, tailing peaks are more common in HILIC operations. The peak shape can occasionally be improved by raising the ionic strength. Adding trifluoroacetic acid (TFA) to the mobile phase leads to improved peak shapes of essential compounds on the silica-based columns by forming ion associates. It should be considered that TFA and high buffer concentrations are not suggested for

HILIC/MS applications since they can inhibit electrospray ionization. Formic acid should always be used instead of TFA [50].

Water is the strongest eluent in HILIC conditions; even a minor increase in water concentration dramatically reduces the retention. The water concentration in the mobile phase should be decreased to 5% or even less to facilitate the separation of poorly retained compounds and promote retention. In several studies, water has been entirely replaced by other polar organic solvents, particularly in polar non-aqueous mobile phases used in conjunction with polar columns. The term given to this technique is "non-aqueous HILIC" (NA-HILIC). Some weakly polar or poorly soluble oligomers that may precipitate in water-containing HILIC mobile phases can be analyzed more effectively with the help of NA-HILIC [3].

1.4.3. Detectors Used in HILIC Chromatography

In HILIC the detectors used are similar to that used in RPLC and NPLC-MS and also UV spectroscopic-based detectors like fluorescence emission and ultraviolet-visible (UV-Vis), Evaporative Light Scattering (ELS). In HILIC separations, fluorescence, and UV absorbance are common; these non-destructive detectors only work with analytes with the proper chromophore groups. For fluorescence emission detection, derivatization is required for analytes lacking fluorophore groups. ELS and electrochemical-based detectors can detect analytes that lack chromophore or fluorophore groups [54].

MS is the most used detector in HILIC and has many applications of separation mode. The main benefits of MS for HILIC are superior direct coupling to MS, adaptability for a wide variety of analytes, improved mobile phase compatibility, and increased detectivity compared to RPLC separations. A mobile phase with a high organic solvent content in HILIC has effective electrospray ionization spraying and desolvation. The environment created by the organic solvent, particularly ACN, is suitable for dissolving analytes and creating tiny charged droplets, facilitating the ionization process [55]. Due to signal suppression in ESI/MS, large acid additions to the mobile phase, particularly TFA, must be controlled [56].

2. Applications

2.1. Food Analysis (Table 1)

2.1.1. Polar Fecal Metabolome Profiling

The fecal metabolome comprises a complex blend of endogenous microbial metabolites and food-derived compounds. Hydrophilic Interaction Liquid Chromatography (HILIC) permits the investigation of polar compounds, which is a helpful substitute for reversed-phase liquid chromatography in the field of metabolomics due to its ability to retain a more significant portion of the polar metabolome. HILIC has frequently been used to analyze hydrophilic compounds in urine and plasma to expand metabolome coverage [18, 57-59]. Sillner *et al.* recently conducted a study utilizing HILIC Ultra High-Pressure Liquid Chromatography Mass Spectrometry (HILIC-UHPLC-MS) for polar fecal metabolome profiling. The study aimed to determine the suitable chromatographic solution for performing non-targeted metabolomics of feces using HILIC-UHPLC-MS. The performance was comprehensively examined using a pooled fecal sample and combinations of 150 metabolites from various families, such as amino acids, amines, indole derivatives, fatty acids, and carbohydrates. Three distinct stationary phases (zwitterionic, amide and unbonded silica) were operated at three pH values (4.6, 6.8 and 9.0), and three salt gradient conditions (5, 5-10 and 5-25 mM ammonium acetate). Amide and zwitterionic stationary phases were also carried out in the same manner at low pH and the highest number of detected standards that increased by increasing the number of salt gradients. The performance of the amide column was better concerning the separation of isomers and better performance at basic pH with more metabolite features in non-targeted analysis, whereas the zwitterionic column used was best concerning the number of detected standards and retention time and characteristics of metabolite across the complete chromatographic run. As a result, the zwitterionic column was the best suited for non-targeted analysis of fecal samples and an investigation of amino acids and carbohydrates.

The primary objective of this study was to obtain a HILIC method that covers the broadest spectra of polar metabolites in fecal matter. The unbonded silica Cortec's HILIC column did not give desired results regarding retention time distribution, precision and a number of recovered standards. Under the studied acidic conditions, BEH Amide and the zwitterionic HILIC Fusion performed similarly, whereas neutral pH produced the poorest results. For primary pH conditions, there was no comparison between the columns because of the shortcoming of pH intolerances. Hence, utilizing the zwitterionic HILIC Fusion column under acidic conditions provides the best conditions for studying the polar fecal metabolome in future non-targeted metabolomics studies [60].

Table 1. HILIC applications in food analysis

S. No	Technique	Analyte	Conditions	LOD/LOQ	Ref.
1	HILIC-UHPLC-MS	Polar fecal metabolome profiling	SP- Zwitterionic, amide, unbounded silica. pH- 4.6, 6.8, 9.0 Salt conditions- 5, 5-10 and 5-25mM ammonium acetate.	—	Sillner *et al.*
2	HILIC-RP-LC-DAD-MS	Natural Phenolics	55-70 min per analysis time.	—	Muller *et al.*
3	HILIC-UHPLC-MS/MS	Amino-acid profiling	Hydrolysed with 6M HCl with 4%(v/v) thioglycolic acid 110°C for 18 hours	LOD-0.002 (valine, serine, leucine, isoleucine) to 0.009 g/100 g (threonine) LOQ- m 0.007 (serine, leucine) to 0.024 g/100 g (threonine, lysine)	Tsochatzis *et al.*
4	HILIC	Food matrices analysis (Bovine milk)	SP- Zwitterionic alkylsulfobetaine Buffer- 6.5mM ammonium acetate MP- Acetonitrile	LOD- 0.05 ng/mL LOQ- 0.25 ng/mL	Marrubini *et al.*
5	HILIC-FLD	Characterization of nutraceuticals and food supplements based on flavanols	SP- Bare silica MP- Acidified acetonitrile and methanol solutions.	—	Canella *et al.*
6	HILIC	Qualitative analysis of phospholipids	0.1-0.2% v/v formic acid, 5-10mM ammonium acetate	—	Parchem *et al.*

2.1.2. Natural Phenolics

Phenolics are a large group of secondary plant metabolites that have gained importance due to their potential health benefits related to antioxidant effects [61]. Plant phenolic analysis is complicated due to its extreme structural diversity [62]. Studying such complex mixtures is done with the help of two-dimensional liquid chromatography (LC × LC) coupled with high-resolution mass spectrometry (HR-MS). However, the combination of Hydrophilic

Interaction Liquid Chromatography (HILIC) and reverse-phase liquid chromatography (RP-LC) is appealing for phenolic analysis because of the attainment of orthogonal group-type separations. Muller *et al.* discovered the application of kinetically optimized online HILIC-RP-LC methods, which are hyphenated with High-Resolution MS for analysis of natural phenolics. A similar approach was later used to determine optimal conditions for peptide RP-LC separation [63, 64]. However, the relative elution strengths of the mobile phases in both dimensions complicate the online hyphenation of HILIC and RP-LC. This, combined with the inherent complexity of method development in LC-LC, impedes the widespread application of HILIC-RP-LC. In this study, a generic HILIC-RP-LC-DAD-MS methodology for phenolic analysis was developed using a dilution of the first-dimension flow and large volume injection in the second dimension, with experimental parameters kinetically optimized to provide maximum performance. The experimental configuration's scope was demonstrated by its application in the analysis of rooibos tea, wine, and grape samples containing a variety of flavonoid and non-flavonoid phenolic classes.

Like most kinetic optimization protocols, this methodology requires the prior selection of stationary and mobile phases in both dimensions to optimize selectivity before kinetic optimization [62]. A predictive kinetic optimization program was used to calculate the chromatographic conditions for the HILIC-RP-LC separations [65]. A predictive optimization protocol with fixed modulation requirements to limit injection band broadening was used in this study to derive kinetically optimized experimental conditions for the online HILIC RP-LC DAD-MS analysis of phenolic compounds that came in the order of 2000 for 55-70 min per analysis time. Using this method, excellent chromatographic performance was obtained, and 149 phenolic compounds were tentatively identified in the investigated samples based on retention data in two dimensions, UV-Vis spectral data and high- and low-collision energy HR-MS data (72 in grape seeds, 32 in rooibos tea, and 45 in wine and grapes) with minimal method development time [62]. Such a relatively high number of unidentified minor constituents from each investigated sample emphasizes the sophistication of natural phenolic mixtures and the importance of developing more powerful analytical tools to explore these. UV contour plots also permit visual evaluation of phenolic profiles, making it easier to examine differences among samples. In conclusion, the kinetically enhanced HILIC-RP-LC DAD-HR-MS technique depending on 1 D effluent dilution and high-volume injection, offers a strong and reasonably fast methodology for precisely screening phenolic content in a variety of natural products [62].

2.1.3. Amino-Acid Profiling

Amino acids are essential nutritional elements in foods in either free states or as protein constituents. Various methods such as Ultra-High-Performance Liquid Chromatography (UHPLC), Hydrophilic Liquid Chromatography (HILIC)-Tandem Mass Spectrometry (MS) methods are created for validation and quantification of the total of 17 amino acids (AA) in the samples of wheat flour after it is hydrolyzed with 6 M HCl in the existence of 4% (v/v) thioglycolic acid that acts as a reducing agent. Tsochatzis *et al.* recently showed validation of HILIC UHPLC-MS/MS method for profiling amino acids in Triticum species of wheat flour. The established methodology is considered a quick and dependable technique for evaluating the AA profile of cereal flours. This procedure was implemented and examined in 10 flour samples of organic or conventionally grown spelt, emmer, and common wheat flours with varying extraction rates (70%, 85% and 100%). All the previous studies allowed us to analyze amino acid profile variations proportionate to their characteristics, such as protein content and gluten index. Substantial variations in the AA profiles of the flours studied were observed. A statistical and multivariate analysis of the AA profiles and quality attributes was performed to identify potential interactions between protein content, amino acids, and quality characteristics. Therefore, details on free AA features and amounts are essential in food science and nutrition studies [66-70].

Recently, a HILIC-MS/MS way of determining free AA or amino acid profiles in a protein-rich food matrix was developed [71]. Developments in hydrophilic interaction chromatography (HILIC) and new analytical columns have opened new avenues for AA profiling via liquid chromatography (LC) [15, 68, 71]. 110°C for 18 hours was applied to reduce (in conjunction with the antioxidant agent) the deterioration of different amino acids while balancing the recovery rates of hydrophobic AA, like valine, isoleucine, and leucine [69, 71, 72]. This temperature was chosen to reduce the possibility of amino-acid cross-reactions with starch. Although it is common to practice in acid hydrolysis to add an antioxidant agent to prevent AA deterioration, in the case of cereal flour analysis, such a protective step had not previously been applied to HILIC-MS. Tryptophan, cysteine, cystine, and asparagine could not be quantified because they deteriorated during hydrolysis due to the severe acidic environment because of the role of ionic interactions in the process, the charge state of the analyte influences the retention time (time that a solute spends in a column) of the sample in HILIC [71, 73]. It's also been demonstrated that proper relation exists not only with the quality attributes of the flour but also with the category of flour [69].

2.1.4. Food Matrices Analysis

Another study was conducted by Marrubini *et al.* that used Hydrophilic Interaction Liquid Chromatography (HILIC) along with various detection techniques to analyze multiple organic components of food, such as nucleobases, nucleosides, nucleotides, uric acid, and creatinine, amino acids and related compounds, choline-related compounds and phospholipids, carbohydrates, artificial sweeteners, and polyphenolic compounds. In addition, the key benefits of HILIC chromatography over other widely utilized methods are covered. Several physiological processes, including the maintenance of the immune response, the growth and differentiation of the gastrointestinal tract, the antioxidant, antiviral, anti-neoplastic, anti-arrhythmic, and anti-seizure activities, are regulated and modulated by the bioactive compounds known as nucleobases (NuBs), nucleosides (NuSs), and nucleotides (NuTs). Several disorders, including gouty arthritis, type 2 diabetes, hypertension, cardiovascular diseases, and kidney diseases, can be prevented by controlling uric acid (UA) and creatinine (CR), physiologic components of human bodily fluids [74].

Organic compounds such as NuBs, NuSs, NuTs, UA, and CR are hydrophilic, polar and non-volatile, and their separations by RPLC techniques are frequently characterized by low selectivity [75]. HILIC has proven to be effective in overcoming many of the constraints and is helpful for the identification and quantification of NuBs, NuSs, NuTs, UA and CR in plant and animal-derived products. In bovine milk, UA and CR were determined with a simple method using isocratic conditions on an amino HILIC column followed by UV detection at 205 nm. The technique described by Zuo *et al.* seems to be an excellent substitute for traditional IP-RPLC and RPLC-based methods [76].

HILIC or electrostatic repulsion liquid chromatography can separate intact proteins, phosphoproteins, and glycoproteins [45]. Broncano *et al.* reported a HILIC-MS/MS analysis for hydrophilic low molecular weight compounds with antioxidant properties in fermented chorizo sausages meat extracts. Using a zwitterionic alkylsulfobetaine stationary phase and various ratios of 6.5 mM ammonium acetate buffer and acetonitrile as the mobile phase system, the target analytes were separated by gradient elution [77].

Porter and Lodge, in their study, developed a technique for detecting specific water-soluble vitamins based on Hydrophilic Interaction Liquid Chromatography (HILIC) coupled with single-quadrupole Mass Spectrometry (MS). The food matrix for this investigation was a readily available vegetable soup. Meta-phosphoric acid combined with a reducing agent and DL-dithiothreitol (DTT) was used to extract vitamins to create the parent compound. HILIC-MS system with an Electrospray-Atmospheric Pressure Ionisation (ES-API) source was used to analyze the extracted vitamins in positive Single Ion Monitoring (SIM) mode. The method demonstrated high linearity and intra- and inter-day reproducibility and can be applied to various plant-based food matrices [78].

2.1.5. Characterization of Nutraceuticals and Food Supplements Based on Flavanols and Related Compounds

The term nutraceutical was proposed by Stephen DeFelice and is composed of two words "nutrient" and "pharmaceutical." Defined as a food or component of nutrition that promotes medical or health benefits, including the ability to treat or prevent disease [79]. Casanella *et al.* presented a paper characterizing distinct types of nutraceuticals using Hydrophilic Interaction Liquid Chromatography (HILIC), employed with fluorescence detection (FLD).

The study's primary objective was to identify flavanols and related compounds, such as condensed tannins, and deal with the development and application of HILIC for the characterization based on flavonol profiling and fingerprinting. Liquid chromatography (LC) is typically used to determine phenolic compounds, especially species related to flavanols. In this regard, several publications have investigated the potential of HILIC separation, generally based on LC-MS/MS platforms, for the analysis of condensed tannins in various food matrices, such as cocoa, [80] blackcurrant and other berries [81, 82] and apple extracts from different varieties [83, 84] In the study, the HILIC mode was created by employing a bare silica column as the stationary phase and acidified acetonitrile and methanol solutions as the mobile phase components.

Chemometric methods were utilized to characterize nutraceuticals and to generate the composition profile and the chromatographic fingerprint as the information source. The results showed that the cranberry extract was abundant in various phenolic compounds, primarily flavanol. Partial Least Squares Discriminant Analysis (PLS-DA) and Principal Component Analysis (PCA) have produced outstanding sample behavior descriptions. Overall, it has been discovered that HILIC-FLD is an excellent method for conducting this application, opening up new opportunities in investigating phenolic compounds.

2.1.6. Qualitative Analysis of Phospholipids (PLs)

Phospholipids (PLs) are critical biomolecules that not only serve as structural basic components and ladders for cell and organelle membranes but also play an essential role in cell biochemistry and physiology. Phospholipids (PL) are polar lipids composed of two fatty acids, a glycerol unit, and a phosphate group esterified to an organic molecule (X) like choline, ethanolamine, or inositol. They are amphiphilic compounds containing lipophilic and hydrophilic moiety, forming building blocks that strengthen biological membranes in animal and plant cells [85]. Parchem *et al.* recently performed a qualitative analysis of phospholipids and their oxidized derivatives with the help of the HILIC chromatographic technique. Various epidemiological studies stated that PLs are identified by vast nutritional values and other health effects like brain function, immune health etc. PLs are of specific importance in lipidomics, which examines both membrane lipid distribution and the composition of lipids about their effects on cell, tissue, and organ function.

Furthermore, polyunsaturated fatty acids (PUFAs) derived from membrane PL are the origin of multiple signaling molecules like eicosanoids, leukotrienes, prostaglandins, and thromboxanes, essential modulators of both inflammation and anti-inflammatory signals [86]. Oxidized PLs (ox PLs) can be produced exogenously and fed to organisms as packaged food or endogenously due to oxidative stress. Oxidized PLs can be present in cells and tissues that can be a biomarker of the origin of various diseases like atherosclerosis or neuroinflammation. That's why appropriate analytical techniques like HILIC that allow an analysis of PL concerning the structure of the hydrophilic group and fatty acids composition have been developed recently. This includes every part of PL analysis, like isolation of lipids, chromatographic separation of PL and their identification.

Liquid chromatography (LC), mainly HILIC, is the most commonly used analytical technique for identifying and quantifying PLs in biological, pharmaceutical, or food specimens, allowing the separation of PL subclasses following their hydrophilic group [87-89]. The duration and degree of unsaturation of FAs and the degree of oxidative modification of FAs existing within the PL structure can be used to differentiate PL classes. [89-91] Columns used in this technique are packed with unmodified silica gel [89, 92], modified diol [93], or amine groups [94] are also used. The mobile phase is composed of a combination of acetonitrile and water [89, 93, 95]. The mobile phase's pH is adjusted by adding 0.1-0.2% v/v formic acid, 5-10 mM ammonium acetate [96]. Time ranges vary for different solutions, i.e., from 60-70 min for gradient elution and 15-20 min for isocratic gradient [97]. Thus, the separation, identification and quantification of phospholipids (PLs) from compounds can be separated with the help of HILIC in a significantly shorter time [87].

2.2. Drug Analysis (Table 2)

2.2.1. Characterization of β-Lactam Drugs in Human Urine

Characterization of *β*-lactam antibiotics in the sample of human urine was done by Solid-Phase Extraction (SPE) before determination by HILIC. A total of 4 cephalosporins, four penicillins, and 1 carbapenem were chosen among all *β*-lactam antibiotics. The stationary phase inside the column silica-linked zwitterionic sulfobetaine group proved helpful in analyzing the above classes of antibiotics (highly hydrophilic) with appropriate peak shape, and resolution acquired by selecting suitable chromatographic conditions. Merlo *et al.* discovered a versatile method for determining *β*-lactam antibiotics drugs in urine samples by solid-phase extraction followed by HILIC-UV. The rapid SPE process developed entirely in urine successfully combined sampling unit clean-up, preventing protein precipitation before extraction and recovery of more than 65% and quantifying limit ranges between 1-4 $\mu g\ mL^{-1}$. Another validation parameter, i.e., selectivity, was determined in blank urine samples in the existence of various non-targeted pharmaceuticals. This procedure was used to analyze samples from four patients receiving *β*-lactam therapy [98]. Antibiotics are considered an essential class of antimicrobials administered orally and parenterally in humans against gram +ve and gram -ve bacteria to protect the body from life-harming infections [99-101]. Because of the greater excretion rate of *β*-lactams in the unprocessed form, urine could be considered an essential and effective biological matrix for testing analysis of these antibiotics in clients [99, 102].

Quantitation was performed using a hydrophilic interaction liquid chromatography (HILIC) stationary phase coupled to UV detection, allowing for a one-run analysis without using costly ion pair reagents or separate columns. Validation parameters like selectivity, linearity, sensitivity, and intra- and inter-day precision [103, 104] were evaluated from human subjects when antibiotics were administered. After analysis, the HILIC results were compared with SPE extract from urine spiked with *β*-lactam antibiotics. The established method is appealing in terms of clean-up, sensitivity, and flexibility for determining *β*-lactams [98].

Table 2. Applications of HILIC in drug/metabolite analysis

S. No	Technique	Analyte	Conditions	LOD/LOQ	Reference
1	SPE-HILIC	β-lactam	limit ranges in between 1-4μg mL^{-1}		Merlo *et al.*
2	HILIC-MS	Metformin	pH 4.8 and acetonitrile (25:75, v/v) at 0.8 mL/min flow rate.	LOD- 0.75ppm LOQ- 2.5ppm	Dousa *et al.*
3	HILIC-LC-MS	Glycyrrhizin from liquorice	12g Teledyne ISCO Gold silica cartridge (20-40 μm)		Liu *et al.*
4	RPLC-HILIC	Metabolic Phenotyping	Polar Stationary Phase. Organic (Aprotic) Mobile Phase- ACN		Rainvelle *et al.*
5	HILIC-MS	Polar analytes like carbohydrates, biogenic amine	Waters BEH HILIC amide column (1.7 μm) with an internal column diameter of 1.0 mm and an overall length of 5 cm		King *et al.*
6	HILIC	Chemometric-assisted separations	MP used are Hydro-organic		Taraji *et al.*

2.2.2. Characterization of Dimethylamine in the Active Pharmaceutical Ingredients and Dosage Forms of Metformin

Metformin (MET) is a rapidly soluble anti-diabetic medication typically used to treat non-insulin-dependent diabetes mellitus [105]. The European Pharmacopoeia describes DMA as an impurity of metformin (impurity F). DMA and other compounds serve as a precursor in forming N-nitrosodimethylamine (NDMA) in water and sewage [106]. The U.S. Environmental Protection Agency classified NDMA as a possible human carcinogen.

A suitable analytical approach based on HILIC coupled with Mass Spectrometric detection (HILIC-MS) was developed by Dousa and Jires, using a straightforward sample preparation. The study's main objective was to create a derivatization-free, accurate, precise, and quick method for measuring DMA in API and dosage forms of metformin. Cortecs HILIC column was used to separate MET, utilizing a mixture of 10 mmol/L ammonium formate adjusted to pH 4.8 and acetonitrile (25:75, v/v) at 0.8 mL/min flow rate. The column temperature for the process was 45°C. The quadrupole mass analyzer was used in a selected ion monitoring mode. The primary advantage of the HILIC-MS method is that an uncomplicated sample pretreatment is required, providing a rapid and sensitive analysis.

According to ICH Q2(R1) guidelines, the analytical method was validated in terms of linearity, detection limit, quantification, selectivity, accuracy, precision, and intermediate precision [107]. The results indicated that the DMA level in API and dosage forms ranged from 16 to 230 ppm. Due to the finding that the method produced good results in all the validation parameters, its frequent use may be considered in the routine quality control in pharmacy.

2.2.3. Separation of Polar Pharmaceuticals

Polar pharmaceuticals are generally difficult to separate by analytical and semi-prep HPLC. A review proposed by Liu *et al.* showed how predictable separations of polar drugs employing low-pressure flash purification in the "HILIC" mode were established.

Flash chromatography is a preparative liquid chromatography technique that provides rapid results by forcing the solvent through the stationary phase column under pressure rather than using slower gravity-based chromatography. There are numerous bioanalytical instances of employing gradient HILIC LC-MS to retain polar analytes, such as urine [108] investigations using a 2D-LC-MS method with RP-LC for drug evaluation in plasma [1] monoamine neurotransmitters [109] and metabonomic/ metabolomic research [110]. In a study, glycyrrhizin (GA) from liquorice was purified using hydrophilic solid phase extraction coupled with RPLC. Although this method enhanced GA retention, HILIC was only used as a crude extraction procedure [111]. A recent way shows the benefits of HILIC over an ion-pair mechanism by describing the use of HILIC for large-scale preparative isolation of capreomycin impurities [112].

Liu *et al.*, in their study, showed the successful development of flash C_{18} purifications from analytical methods, so it was required to determine if the same procedure could be used for HILIC silica flash purifications [113]. A polar test mix was selected for the process using small molecule polar compounds, nicotinamide, amino acid tryptophan, small essential molecules caffeine, procainamide, and nucleic bases purine and cytosine. A 12 g Teledyne ISCO Gold silica cartridge (20-40 μm) was utilized, and the total amount loaded was 20 mg. The Gold silica flash columns can be used again, providing the same separation even after 10 injections. However, other phases apart from silica can also be utilized for HILIC separations.

Hence, the demonstrated method provided ease of use for medicinal chemists, reduced capital and consumable costs, and decreased fraction quantities. The primary aspect is that instead of having to send materials for HILIC semi-prep, normal phase, or SFC separations, chemists can now purify highly polar compounds using HILIC flash, which is easily accessible at their fume hood.

2.2.4. Metabolic Phenotyping

Several LC separations are undertaken using reversed-phase (RP) methods, with hydrophilic interaction liquid chromatography (HILIC) being a common choice for polar compounds that are not very well provided by the RP mode. The use of ultra-high performance LC [UHPLC], HILIC is rapidly replacing high-performance LC (HPLC) in metabonomic/metabolomic applications, as its advantages are visible as essential for quick, high-resolution sample analysis. Rainville *et al.* showed advances in Hydrophilic Interaction Liquid Chromatography (HILIC) coupled with Mass Spectrometry for metabolic phenotyping [114]. Using metabolites as biomarkers for disease diagnosis and understanding, diagnostic tools, drug safety evaluation, and basic physiology experiments are not novel [115]. LC-MS uses have been employed in reverse-phase (RP) separations, also with Hydrophilic Interaction Liquid Chromatography (HILIC) is getting increasingly prominent as the limitations of RPLC in evaluating the polar fraction of the metabolome became clear [116]. Moreover, there is no optimal solution when analyzing a complex mixture with a wide range of analyte polarities, to choose the best balance that provides the most coverage of components in the sample while remaining compatible with the detector used is the requirement. Due to this reason, most of the metabolic phenotyping experiments are performed with RPLC-HILIC.

HILIC uses a polar stationary phase and a large amount of organic (aprotic) mobile phase, such as acetonitrile, like the eluent (modified with water as the solid eluting solvent), becoming more popular for polar metabolites. In this technique, more hydrophilic compounds will get attracted to the stationary phase and will elute later and more lipophilic compounds will not get attracted to the stationary phase and will elute first. Retained polar compounds inside the stationary phase will only be eluted when polar conditions are increased. Aqueous samples have good compatibility with the mobile phase in HILIC. HILIC is used in conjunction with MS/MS in the so-called targeted metabolic profiling mode to develop multi-analyte examination [110, 117]. Tremendous chromatographic scientific progress has occurred over the last few years, improving data quality, density, sensitivity, and compatibility with sample solvents [114].

2.2.5. Analysis of Polar Metabolites in Urine

Large-scale preclinical metabolism/toxicological [118-120], clinical, or epidemiological [121, 122] investigations using metabolic phenotyping have revealed new details about health, toxicity, and illness biology. The biological matrices used in metabolomic research contain various compounds, resulting in different chemical characteristics. It can be challenging to completely cover all of these molecules with RP separations, particularly for polar analytes (e.g., carbohydrates, organic acids and biogenic amines) that are weakly retained. Hydrophilic interaction liquid chromatography (HILIC) has made it possible to analyze a variety of polar molecules [123]. As a result, it is now routine practice to use HILIC for the dual analysis of samples like urine to perform the metabolite analysis of biofluids [124, 125]. The fusion of HILIC and RPLC has started to be applied as a standard practice [57, 58, 126, 127].

King *et al.*, in a review, described using HILIC-MS alone or in combination with IMS-enabled MS data acquisition (HILIC-IMS-MS) to analyze polar metabolites in urine using a method somewhat similar to the reversed-phase Rapid Microbore Metabolic Profiling (RAMMP) method [128]. The urine samples required for the stated study were obtained from male Sprague-Dawley rats dosed intravenously [129]. A Waters BEH HILIC amide column (1.7 μm) with an internal column diameter of 1.0 mm and an overall length of 5 cm was used for the HILIC separation of the polar metabolites. A gradient elution procedure was employed for the purpose.

According to chromatographic theory, peak width is directly related to column length, so halving the size should produce narrower peaks. The results showed that the Rapid HILIC method had narrower peaks than the conventional HILIC method. In contrast to the traditional UPLC-HILIC/MS method, the newly formed rapid HILIC process offered a four-fold reduction in analysis time, a 75% reduction in solvent use, and an 18-fold reduction in sample consumption over a conventional HILIC UPLC method—all while maintaining the ability to detect important features that distinguish between various sample groups.

2.2.6. Chemometric-Assisted (Computer-Aided) Separation of Metabolomes

Theoretical research in computer-aided modeling tools, particularly predictive, quantitative structure, and QSRR modeling methods, have researchers' interest, and these strategies vastly aid project development. Computer-aided modeling tools recognize the retention mechanism, classify HILIC stationary phases, optimize various chromatographic conditions, and describe the interaction effects of chromatographic aspects in HILIC separations. Taraji *et al.* recently developed a Chemometric-assisted method in Hydrophilic Interaction Liquid Chromatography. Creating robust predictive quantitative structure retention relationship (QSRR) models will doubtlessly enable so many applications of this chromatographic mode in a broader range of 41 areas of research, significantly reducing experimental cost and time. The availability of a diverse range of stationary phases in hydrophilic interaction liquid chromatography (HILIC) increases selectivity and retention [130-132]. Those same methodologies for modeling analyte retention are known asQSRR modeling [133-135].

The mobile phase used in HILIC is hydro-organic (both hydrophilic and lipophilic) in nature that allows good hyphenation with Mass Spectrometry (MS) which allows good retention of very polar analytes [136]. Another use of HILIC columns is separating hydrophilic solutes and metabolomes [47, 50]. Overall, enhancing the percentage of organic solvent (ACN) inside the mobile phase improves the retention of hydrophilic analytes on various types of HILIC columns. Therefore when the ratio of Acetonitrile (ACN) was between 60-90%, the retention time of the analyte in the stationary phase was increased by increasing acetonitrile concentration and when acetonitrile content was in the range of 5-35%, there was no change in the retention time of any analyte [50]. QSRRs were seen to be computationally efficient, capable of handling and interpreting large data sets, robust, and accurate in retention prediction. Chemometrics approaches such as QSRRs have proven to be useful for functional formulation development in HILIC when combined with the design of experiments techniques in a QbD workflow [135].

2.3. Agricultural Analysis (Table 3)

2.3.1. Residue Analysis of Dithiocarbamate Fungicides

Hydrophilic Interaction Liquid Chromatography (HILIC) is the primary method used for the selective detection of dithiocarbamate (DTC) fungicides, which have been frequently used in agriculture for many decades. Crnogorac *et al.* performed a simultaneous estimation of DTC fungicide residues on fruits and vegetables using a zwitterionic polymer-based Hydrophilic Interaction Liquid Chromatography column (ZIC-pHILIC) column coupled with tandem mass spectrometry (MS/MS) [137]. The detection was carried out by mass spectrometry using a single quadrupole mass spectrometer connected to an electrospray ionization interface operating in the negative mode. Generally, carbon disulfide (CS_2) is used to express the maximum residue limits for DTCs. Therefore, techniques for residue analysis are based on the acid hydrolysis evolving CS_2 from all DTCs [138].

Table 3. Applications of HILIC in agriculture analysis

S. No	Technique	Analyte	Condition	LOD	Reference
1	HILIC-MS/MS	dithiocarbamate fungicides H-DMD H_2-EBD H_2-PBD	mobile phase containing 95:5 (v/v) (water:0.1% acetonitrile).	0.05μg/kg 1.2μg/kg 0.10μg/kg	Crnogorac *et al.*
2	HILIC-ELSD	Glufosinate herbicides Tea Leaves Soil	mobile phase 95:5 (v/v) (water:0.1% acetonitrile).	0.01mg/kg 0,01mg/kg	Janaki *et al.*

In 2007, the ZIC-pHILIC was introduced, which opened up several opportunities for residue analysis of DTCs in fruits and vegetables [137, 139]. Pesticides in the "maneb group" (zineb, maneb, mancozeb, propineb, and metiram), which are fungicidal dithiocarbamates (DTCs), were the most frequently found pesticides and the ones that exceeded Maximum Residue Limits (MRLs) the most frequently in many different nations [140]. Dimethyldithiocarbamates (DMDs), ethylenebis (dithiocarbamates) (EBDs), and propylenebis (dithiocarbamates) (PBDs) were the three DTC subclasses for which residue analysis was performed [138].

2.3.2. Assessment of Glufosinate Dissipation in Soil and Its Residues

Due to its ionic and highly water-soluble nature, separating and detecting glufosinate from various chromatography matrices from lower than Minimum Residual Levels (MRL) is challenging. A fast and sensitive method has been developed for determining glufosinate in soil and water with the help of Hydrophilic Interaction Liquid Chromatography (HILIC) coupled with an Evaporative Light Scattering Detector (ELSD). Janaki *et al.* recently assessed glufosinate dissipation behavior in soil and its terminal residues with the help of Hydrophilic Interaction Liquid Chromatography (HILIC). The method has been validated by using a spiked blank matrix, and the average glufosinate recovery varied from 78.4 to 90.5%, with a replicate standard deviation (RSD) ranging from 1.00 to 2.39%. Further validation was done with exact samples from tea field experiments performed with varying glufosinate rates.

Glufosinate is a naturally present phytotoxin referred to as phosphinothricin that is formed by soil streptomyces species used to manage a wide variety of weeds. [141] Agilent HILIC column of diameter 5 μm was used. Further, the glufosinate was eluted with mobile phase containing 95:5 (v/v) (water:0.1% acetonitrile). The flow rate was maintained from 0.25 to 1 mL min^{-1} and the temperature was maintained with the help of a thermostat at 50°C with a sample injection volume of 10 μL. This led to desired retention and brilliant separation of highly polar glufosinate. The HILIC separation and optimized ELSD conditions were found to be promising for separating and detecting glufosinate in all samples without involvement from several other co-extractants using HPLC [142].

2.4. Adulterant Analysis (Table 4)

2.4.1. Detection of Adulterant in Fresh-Milk Sample

Because of the variability of urea levels in milk, adulteration with exogenous urea was previously difficult to detect. So that's why nowadays, $^{15}N_2$-Urea is taken as a standard. This method detects adulteration in fresh milk containing compounds like melamine, ammeline, and cyanuric acid that contain nitrogen as a functional group. By examining the contaminants like biuret and triuret (which include synthetic urea up to 1%), we can analyze the adulteration in milk. Abernethy *et al.* rapidly detected economic adulterants in fresh milk by Hydrophilic Interaction Liquid chromatography. To be economically viable, he estimated that 90-4000 ppm of the preceding adulterants would need to be added to fresh milk. A lower detection threshold, more significant than 0.5 ppm, is desirable for biuret, and the specificity for biuret and triuret can be improved by adding lithium post-column to create lithium adducts under electrospray ionization. The nitrogen concentration may vary due to natural differences in the milk and the collection procedure or, more unusually, due to intentional adulteration of the milk with specific nitrogen-containing compounds [143, 144]. The sample is managed using a two-step solvent precipitation method in a 96-well plate feature, and the hydrophilic interaction liquid chromatography (HILIC) employs a rapid gradient in 1.2 min.

Even though biuret can be spotted at 1-2 ppm using HILIC, the signal-to-noise ratio is low, and that can be enhanced 4-5-folds by adding lithium to the mobile phase post-column to form lithium complexes. The method employed a HILIC column (Phenomenex Kinetex 50 mm × 2.1 mm, 2.6 m core-shell; Phenomenex, Torrance, CA, USA) with mobile phases containing 0.1 percent ammonium formate in water and 0.1 percent formic acid in acetonitrile (ACN) and run at a flow rate of 1 mL m^{-1}. Thus detection of Nitrogen-based adulteration in fresh milk was done by HILIC Chromatography [145].

Table 4. Analysis of adulterants through HILIC

S. No	Technique	Analyte	Condition	LOD	Reference
1	HILIC	Fresh milk sample	containing 0.1 percent ammonium formate in water and 0.1 percent formic acid in acetonitrile (ACN) and run at a flow rate of 1 mL m^{-1}.	More than 1ppm	Abernethy *et al.*
2	HILIC-MS/MS	Nitrogen-rich food samples. Cyanuric acid Biuret Melamine	Mobile Phase A- 10 mM ammonium acetate in water Mobile Phase B- 10mM ammonium acetate in 97:3 (v/v) acetonitrile/water	0.2mg/kg 4mg/kg 0.1mg/kg	Draher *et al.*

2.4.2 Nitrogen-Rich Adulterant Analysis Taken from Food Samples

A technique for analysis of various nitrogen-rich economic adulterants like urea, dicyandiamide, amidinourea, and melamine in various milk and soy-derived food products has been developed for daily use. The samples were diluted tenfold with distilled water before being precipitated with 2 percent formic acid and acetonitrile (ACN). The extracted amount of sample was analyzed with the help of HILIC-MS/MS concerning both positive and negative modes. Draher *et al.* validated a rapid method for analyzing nitrogen-rich adulterants in food products using HILIC-tandem mass spectrometry [146]. In multi-day validation experiments, the average accuracies and method detection limits (MDL) for all analytes in complete milk powder were 82-101 percent, 6-13 percent, and 0.1 mg/kg-7 mg/kg, respectively. Analyte retention times in spiked matrix controls were within ±0.06 min of the average retention times of related analytes in calibration standards.

The current effort is focused on developing a susceptible and easy-to-use HILIC chromatography-based MS/MS method for quantifying economic adulterants in milk-derived ingredients. A Waters Acquity BEH HILIC (2.1 mm × 150 mm × 1.7 μm) UPLC column was taken to separate chromatographic samples. Two mobile phase compositions were used as follows- (1) Mobile Phase A-10 mM ammonium acetate in water, (2) Mobile Phase B- 10 mM ammonium acetate in 97:3 (v/v) acetonitrile/water. The flow rate of the pump was maintained at 700 μL/min. The methods selectivity was checked by checking three single lots of blank samples for any interference at retention time. A LC/MS technique was developed to analyze nine different adulterants in five of the protein ingredients derived from milk and plant protein in less than 7 minutes, more precisely with HILIC-MS/MS [146].

2.5. Field of Proteomics (Table 5)

2.5.1. Amino Acid Analysis for Protein Quantification

Techniques that are used to calculate the protein content from biological samples are most of the times not correct due to errors. Losses cause errors during protein preparation and incorrect detection of interfering compounds or elements. Analysis of Amino Acids includes steps of the chromatographic process that are used to measure protein levels, help get exact compositional information, and acts as a complementary approach to peptide mass fingerprinting [147]. Kambhampati *et al.* found accurate and efficient amino acid analysis for protein quantification using Hydrophilic Interaction Liquid Chromatography coupled with tandem mass spectrometry. Although unstable derivative products are harmful and expensive, incomplete reactions and insufficient chromatographic separations all impede precise protein quantitation using AAA. HILIC columns split components between a significantly hydrophobic mobile phase and a sufficiently hydrophilic stationary phase to retain partially water-soluble compounds [148]. HILIC-based strategies have successfully resolved both proteinogenic and non-proteinogenic amino acids [70, 149, 150] in animal and plant matrices [71, 151-154]. HILIC can be used to analyze protein content without any addition of chromophore.

Table 5. Applications of HILIC in proteomics

S. No	Technique	Analyte	Condition	LOD	Reference
1	HILIC-MS/MS	Amino Acids 1. Alanine 2. Arginine 3. Cytosine	Infinity Lab Poroshell 120 Z-HILIC column.	11.62 pmol 0.83 pmol 0.20 pmol	Kambhampati *et al.*
2	SPE-HILIC	Proteins	10% SDS or 8 M urea.	----	Ye *et al.*
3	HILIC-MS	Histones, a membrane protein.	High concentrations of TFA in the mobile phase (0.1% v/v),	----	Gargano *et al.*

HILIC-MS/MS is utilized to see free amino acids and is tested with many hydrolysis parts to estimate the protein level. All twenty amino acids were resolved and quantified using a widely viable mix of labeled amino acids, quantitatively determining protein levels. This method provides a reliable and effective strategy that could be expanded to detect non-proteinogenic and modified amino acids [70, 149, 150]. Similarly, for the determination of combined protein concentration, isotopic structures are used by combining HILIC LC-MS/MS with High-Resolution Mass Spectrometry (HRMS) [155, 156].

2.5.2. Integrated Proteomics Sample Fractionation

Hydrophilic Interaction Liquid Chromatography (HILIC) has become a vital and daily performed approach for proteome profiling. There are four categories by which sample separation methods by protein trapping mechanisms can be done- Ion exchange chromatography, Hydrophilic Interaction Liquid Chromatography (HILIC), size exclusion chromatography and reverse phase chromatography. Ye *et al.* researched integrated proteomics sample preparation and fractionation by the Solid-Phase extraction (SPE)-HILIC technique. The SPE technique is carried out in a single pipe with hydrophilic paramagnetic beads capable of immobilizing proteins and peptides in hydrophilic interaction chromatography (HILIC) mode.

Even though the HILIC mode has relatively less retention for analytes, particularly in the case of high buffering salts, this retention performance is excellent [157]. The captured proteins and peptides are then cleaned efficiently on a magnetic rack with near to 10% SDS or 8 M urea. Trypsin is then used to digest the eluted proteins in 5 μL of digestion buffer overnight [158]. This leads to the separation of proteins from their analyte with the help of SPE-HILIC. [16]

2.5.3. Top-Down Proteomics

In the last few years, there has been satisfactory progress in Top-Down Proteomics for chromatographic methods with mass spectrometry (MS) which can differentiate various proteins. Hydrophilic Interaction Liquid Chromatography (HILIC) shows promising results for the characterization of glycoforms of different proteins. HILIC can determine many proteins with orthogonal selectivity compared to reversed-phase LC (RPLC). Gargano *et al.* discovered a capillary HILIC-MS for sensitive Top-Down Proteomics. Furthermore, HILIC-MS sensitivity is reduced due to ion suppression from trifluoroacetic acid (TFA), which is frequently used as an ion-pair agent to enhance the selectivity and efficacy of glycoprotein analysis. The HILIC-MS method solves these issues.

The two most common types of stationary phases are weak ion exchange [159] and neutral, i.e., polyhydroxylated or amide functionalized) materials [31, 160]. Histones, membrane proteins, intact and IdeS-digested monoclonal antibodies, biopharmaceuticals, and neoglycoproteins have all been studied using HILIC-MS [159, 161-164]. Given its alternative selectivity and high peak capacity, HILIC can be used as a pre-fractionation technique to minimize sample complexity (e.g., in top-down proteomics analysis of complex cell lysates) [165]. Thus capillary HILIC-MS method can be used to separate complex mixtures like cell lysates using broad-range mobile-phase gradients (e.g., from 10 to 40 percent water) [16].

2.6. Field of Impurity Profiling (Table 6)

2.6.1. Characterization of Antisense Oligonucleotide Impurities

Impurity profiles of oligonucleotides are commonly determined using ion-pairing reversed-phase liquid chromatography (IPRP) with a mobile phase containing triethylamine (TEA) and hexafluoro-2-propanol (HFIP). But TEA has ion-suppressing effects that hamper Mass Spectrometry (MS) and on the other hand, due to the corrosive nature of HFIP, it can affect the robustness of MS. Anion Exchange Chromatography (AEC) is a perpendicular mode to IPRP. However, it cannot be directly coupled with MS. Goyon *et al.* recently with the help of AEX-HILIC and IPRP-HILIC characterized antisense oligonucleotide impurities. The HILIC mobile phase comprised nearly 25 mM ammonium acetate and allowed the detection of contaminants at concentrations as low as 0.3 percent. The 2D HILIC column enabled the online design of AEX and IPRP modes and additional impurity separation.

Table 6. Applications of HILIC in impurity profiling

S. No	Technique	Analyte	Condition	LOD	Reference
1	AEX-HILIC and IPRP-HILIC	Oligonucleotides	The flow rate was set at 0.40 mL/min and then increased to 0.8 mL/min after 8 minutes	Above 3	Goyon *et al.*
2	HILIC	Bisoprolol	Kinetex HILIC 100Å (100 mm x 4.5 mm, 2.6 μm particle size) made up of bare silica column. MP-acetonitrile with ammonium acetate buffer.	—	Farmaciju *et al.*

When directing the HILIC column to the MS instrument, the flow rate was set at 0.40 mL/min and then increased to 0.8 mL/min after 8 minutes when the 2D flow was diverted to the waste. The volume of the HILIC column was improved by using a 3.0 mm I.D × 150 column. This 2D column separates the impurities based on their polarity, and the dissolved impurities were simultaneously identified by MS [166]. HILIC mobile phase is essential because Antisense Oligonucleotide (ASO) at acidic pH did not elute, but at basic and neutral pH, separation of ASO was possible [167].

2.6.2. Analysis of Bisoprolol and Its Impurities

This represents the robustness of Hydrophilic Interaction Liquid Chromatography (HILIC) for analyzing bisoprolol and its impurities. Bisoprolol is a medicine used to treat High Blood Pressure (Hypertension). Acetonitrile (ACN) is used as a mobile phase, and pH and buffer concentration are maintained in the water phase. Farmaciju *et al.* analyzed bisoprolol and its impurities using the HILIC method. The column used was Kinetex HILIC 100Å (100 mm x 4.5 mm, 2.6 μm particle size) made up of bare silica column, whereas the mobile phase consisted of acetonitrile with ammonium acetate buffer and glacial acetic acid [12, 50].

Retention of HILIC depends on these factors- 1) Surface characteristics of the stationary phase 2) Structural properties of compounds that are analyzed 3) mobile phase composition. In HILIC systems, less polar mobile phases have lower elution strength, prolonging the retention time of polar analytes. Thus, analytes in the sample can be separated based on their polarities. The statistical suitability of the theoretical models describing the retention behavior of the analyzed substances is confirmed based on their polarity.

Conclusion

HILIC has shown tremendous applications in various fields. The term "HILIC" can refer to a wide range of methods that combine polar stationary phases with aqueous-organic (or polar organic) mobile phases. These methods typically exhibit normal-phase retention behavior, which can be somewhat influenced by ion exchange (IEX), ion repulsion, size-exclusion (SEC), ion-pairing (IP), or reversed-phase (RP) mechanism contributions. The last few decades have brought spectacular growth in this field, undoubtedly due to the increasing significance of separating low-molecular or high-molecular highly polar compounds, which is challenging to achieve in conventional HPLC modes. As described in this chapter, HILIC has been employed in various fields but still has some limitations. The main drawback of HILIC can be considered as its heavy dependency on the aprotic solvent acetonitrile.

Moreover, it can be assumed that HILIC is a less adaptable technology given the highly regarded and well-understood method development linked with reversed-phase liquid chromatography. Further, advancements have been made in HILIC for glycopeptide enrichment and separation of glycans. Future use and popularity analyses of HILIC are likely to rise due to ongoing research into new HILIC techniques coupled with more potent detection systems.

References

[1] Isokawa, M., T. Kanamori, T. Funatsu, M. Tsunoda, Recent advances in hydrophilic interaction chromatography for quantitative analysis of endogenous and pharmaceutical compounds in plasma samples, *Bioanalysis,* 6 (2014) 2421-2439.

[2] Qing, G., J. Yan, X. He, X. Li, X. Liang, Recent advances in hydrophilic interaction liquid interaction chromatography materials for glycopeptide enrichment and glycan separation, *TrAC Trends in Analytical Chemistry,* 124 (2020) 115570.

[3] Bicker, W., J. Wu, M. Lämmerhofer, W. Lindner, Hydrophilic interaction chromatography in nonaqueous elution mode for separation of hydrophilic analytes on silica-based packings with noncharged polar bondings, *Journal of separation science,* 31 (2008) 2971-2987.

[4] Soukup, J., P. Jandera, Comparison of nonaqueous hydrophilic interaction chromatography with aqueous normal-phase chromatography on hydrosilated silica-based stationary phases, *Journal of separation science,* 36 (2013) 2753-2759.

[5] Guo, Y. Recent progress in the fundamental understanding of hydrophilic interaction chromatography (HILIC), *Analyst,* 140 (2015) 6452-6466.

[6] Cortés Valle, S. *Characterization of HILIC systems: underivatized silica as stationary phase,* (2020).

[7] Li, R., Y. Guo, Q. Yuan, Recent applications of hydrophilic interaction chromatography in environmental analysis, *Journal of liquid chromatography & related technologies,* 34 (2011) 1112-1132.

[8] Song, Q., Y. Zhao, X. Chen, J. Li, P. Li, Y. Jiang, Y. Wang, Y. Song, P. Tu, New instrumentation for large-scale quantitative analysis of components spanning a wide polarity range by column-switching hydrophilic interaction chromatography-turbulent flow chromatography-reversed phase liquid chromatography-tandem mass spectrometry, *RSC advances,* 7 (2017) 31838-31849.

[9] Gama, M.R., C.B. Bottoli, Hydrophilic Interaction Liquid Chromatography of Small Molecules, Encyclopedia of Analytical Chemistry: Applications, *Theory and Instrumentation,* (2006) 1-23.

[10] Zhang, Q., F.Q. Yang, L. Ge, Y.J. Hu, Z.N. Xia, Recent applications of hydrophilic interaction liquid chromatography in pharmaceutical analysis, *Journal of separation science,* 40 (2017) 49-80.

[11] Peris-García, E., M.J. Ruiz-Angel, J.J. Baeza-Baeza, M.C. García-Alvarez-Coque, Comparison of the fitting performance of retention models and elution strength behaviour in hydrophilic-interaction and reversed-phase liquid chromatography, *Separations,* 8 (2021) 54.

[12] Greco, G., T. Letzel, Main interactions and influences of the chromatographic parameters in HILIC separations, *Journal of Chromatographic Science,* 51 (2013) 684-693.

[13] Subirats, X., M.H. Abraham, M. Rosés, Characterization of hydrophilic interaction liquid chromatography retention by a linear free energy relationship. Comparison to reversed-and normal-phase retentions, *Analytica chimica acta,* 1092 (2019) 132-143.

[14] Guo, Y., S. Gaiki, Retention behavior of small polar compounds on polar stationary phases in hydrophilic interaction chromatography, *Journal of Chromatography A,* 1074 (2005) 71-80.

[15] Bernal, J., A.M. Ares, J. Pól, S.K. Wiedmer, Hydrophilic interaction liquid chromatography in food analysis, *Journal of Chromatography A,* 1218 (2011) 7438-7452.

[16] Gargano, A.F., L.S. Roca, R.T. Fellers, M. Bocxe, E. Domínguez-Vega, G.W. Somsen, Capillary HILIC-MS: a new tool for sensitive top-down proteomics, *Analytical Chemistry,* 90 (2018) 6601-6609.

[17] Easter, R., C. Barry, J. Caruso, P. Limbach, Separation and identification of phosphorothioate oligonucleotides by HILIC-ESIMS, *Analytical Methods,* 5 (2013) 2657-2659.

[18] Virgiliou, C., I. Sampsonidis, H.G. Gika, N. Raikos, G.A. Theodoridis, Development and validation of a HILIC-MS/MS multitargeted method for metabolomics applications, *Electrophoresis,* 36 (2015) 2215-2225.

[19] Gilar, M., P. Olivova, A.E. Daly, J.C. Gebler, Orthogonality of separation in two-dimensional liquid chromatography, *Analytical chemistry,* 77 (2005) 6426-6434.

[20] Appiah-Amponsah, E., K. Owusu-Sarfo, G.N. Gowda, T. Ye, D. Raftery, Combining hydrophilic interaction chromatography (HILIC) and isotope tagging for off-Line LC-NMR applications in metabolite analysis, *Metabolites,* 3 (2013) 575-591.

[21] Ramdzan, A.N., L. Barreiros, M.I.G. Almeida, S.D. Kolev, M.A. Segundo, Determination of salivary cotinine through solid phase extraction using a bead-injection lab-on-valve approach hyphenated to hydrophilic interaction liquid chromatography, *Journal of Chromatography A,* 1429 (2016) 284-291.

[22] Linden, J.C., C.L. Lawhead, Liquid chromatography of saccharides, *Journal of chromatography A,* 105 (1975) 125-133.

[23] Rubinstein, M. Preparative high-performance liquid partition chromatography of proteins, *Analytical Biochemistry,* 98 (1979) 1-7.

[24] Hemström, P., K. Irgum, Hydrophilic interaction chromatography, *Journal of separation science,* 29 (2006) 1784-1821.

[25] Tolstikov, V.V., O. Fiehn, Analysis of highly polar compounds of plant origin: combination of hydrophilic interaction chromatography and electrospray ion trap mass spectrometry, *Analytical biochemistry,* 301 (2002) 298-307.

[26] Schlichtherle-Cerny, H., M. Affolter, C. Cerny, Hydrophilic interaction liquid chromatography coupled to electrospray mass spectrometry of small polar compounds in food analysis, *Analytical Chemistry,* 75 (2003) 2349-2354.

[27] Guo, Y., S. Gaiki, Retention and selectivity of stationary phases for hydrophilic interaction chromatography, *Journal of Chromatography A,* 1218 (2011) 5920-5938.

[28] Dell'Aversano, C., G.K. Eaglesham, M.A. Quilliam, Analysis of cyanobacterial toxins by hydrophilic interaction liquid chromatography–mass spectrometry, *Journal of Chromatography A,* 1028 (2004) 155-164.

[29] Dallet, P., L. Labat, E. Kummer, J. Dubost, Determination of urea, allantoin and lysine pyroglutamate in cosmetic samples by hydrophilic interaction chromatography, Journal of Chromatography B: *Biomedical Sciences and Applications,* 742 (2000) 447-452.

[30] Ke, S.-H., G.S. Coombs, K. Tachias, D.R. Corey, E.L. Madison, Optimal subsite occupancy and design of a selective inhibitor of urokinase, *Journal of Biological Chemistry*, 272 (1997) 20456-20462.

[31] Tetaz, T., S. Detzner, A. Friedlein, B. Molitor, J.-L. Mary, Hydrophilic interaction chromatography of intact, soluble proteins, *Journal of Chromatography A,* 1218 (2011) 5892-5896.

[32] McClintic, C., D.M. Remick, J.A. Peterson, D.S. Risley, Novel method for the determination of piperazine in pharmaceutical drug substances using hydrophilic interaction chromatography and evaporative light scattering detection, *Journal of liquid chromatography & related technologies,* 26 (2003) 3093-3104.

[33] Wang, X., W. Li, H.T. Rasmussen, Orthogonal method development using hydrophilic interaction chromatography and reversed-phase high-performance liquid chromatography for the determination of pharmaceuticals and impurities, *Journal of Chromatography A,* 1083 (2005) 58-62.

[34] Karatapanis, A.E., Y.C. Fiamegos, C.D. Stalikas, HILIC separation and quantitation of water-soluble vitamins using diol column, *Journal of separation science,* 32 (2009) 909-917.

[35] Jandera, P., T. Hájek, Utilization of dual retention mechanism on columns with bonded PEG and diol stationary phases for adjusting the separation selectivity of phenolic and flavone natural antioxidants, *Journal of separation science,* 32 (2009) 3603-3619.

[36] Wu, J., W. Bicker, W. Lindner, Separation properties of novel and commercial polar stationary phases in hydrophilic interaction and reversed-phase liquid chromatography mode, *Journal of separation science,* 31 (2008) 1492-1503.

[37] Mitchell, C.R., D.W. Armstrong, Cyclodextrin-based chiral stationary phases for liquid chromatography, *Chiral Separations,* (2004) 61-112.

[38] Wang, C., C. Jiang, D.W. Armstrong, Considerations on HILIC and polar organic solvent-based separations: Use of cyclodextrin and macrocyclic glycopetide stationary phases, *Journal of separation science,* 31 (2008) 1980-1990.

[39] Tang, Y., J. Zukowski, D. Armstrong, Investigation on enantiomeric separations of fluorenylmethoxycarbonyl amino acids and peptides by high-performance liquid chromatography using native cyclodextrins as chiral stationary phases, *Journal of Chromatography A,* 743 (1996) 261-271.

[40] Risley, D.S., M.A. Strege, Chiral separations of polar compounds by hydrophilic interaction chromatography with evaporative light scattering detection, *Analytical chemistry,* 72 (2000) 1736-1739.

[41] Dejaegher, B., Y. Vander Heyden, HILIC methods in pharmaceutical analysis, *Journal of separation science*, 33 (2010) 698-715.

[42] Guo, Y., A. Huang, A HILIC method for the analysis of tromethamine as the counter ion in an investigational pharmaceutical salt, *Journal of pharmaceutical and biomedical analysis,* 31 (2003) 1191-1201.

[43] Oyler, A.R., B.L. Armstrong, J.Y. Cha, M.X. Zhou, Q. Yang, R.I. Robinson, R. Dunphy, D.J. Burinsky, Hydrophilic interaction chromatography on amino-silica phases complements reversed-phase high-performance liquid chromatography and capillary electrophoresis for peptide analysis, *Journal of chromatography A,* 724 (1996) 378-383.

[44] de Person, M., A. Hazotte, C. Elfakir, M. Lafosse, Development and validation of a hydrophilic interaction chromatography-mass spectrometry assay for taurine and methionine in matrices rich in carbohydrates, *Journal of chromatography A,* 1081 (2005) 174-181.

[45] Alpert, A.J. Electrostatic repulsion hydrophilic interaction chromatography for isocratic separation of charged solutes and selective isolation of phosphopeptides, *Analytical chemistry,* 80 (2008) 62-76.

[46] Chirita, R.-I., C. West, A.-L. Finaru, C. Elfakir, Approach to hydrophilic interaction chromatography column selection: application to neurotransmitters analysis, *Journal of chromatography A,* 1217 (2010) 3091-3104.

[47] Dejaegher, B., D. Mangelings, Y. Vander Heyden, Method development for HILIC assays, Journal of separation science, 31 (2008) 1438-1448.

[48] Takegawa, Y., K. Deguchi, H. Ito, T. Keira, H. Nakagawa, S.I. Nishimura, Simple separation of isomeric sialylated N-glycopeptides by a zwitterionic type of hydrophilic interaction chromatography, *Journal of separation science,* 29 (2006) 2533-2540.

[49] Onorato, J.M., R. Langish, A. Bellamine, P. Shipkova, Applications of HILIC for targeted and non-targeted LC/MS analyses in drug discovery, *Journal of separation science,* 33 (2010) 923-929.

[50] Jandera, P. Stationary and mobile phases in hydrophilic interaction chromatography: a review, *Analytica chimica acta,* 692 (2011) 1-25.

[51] Quiming, N.S., N.L. Denola, A.B. Soliev, Y. Saito, K. Jinno, Retention behavior of ginsenosides on a poly (vinyl alcohol)-bonded stationary phase in hydrophilic interaction chromatography, *Analytical and bioanalytical chemistry,* 389 (2007) 1477-1488.

[52] Fountain, K.J., J. Xu, D.M. Diehl, D. Morrison, Influence of stationary phase chemistry and mobile-phase composition on retention, selectivity, and MS response

in hydrophilic interaction chromatography, *Journal of separation science,* 33 (2010) 740-751.
[53] dos Santos Pereira, A., A.J. Girón, E. Admasu, P. Sandra, Green hydrophilic interaction chromatography using ethanol–water–carbon dioxide mixtures, *Journal of separation science,* 33 (2010) 834-837.
[54] Cintrón, J.M., D.S. Risley, Hydrophilic interaction chromatography with aerosol-based detectors (ELSD, CAD, NQAD) for polar compounds lacking a UV chromophore in an intravenous formulation, *Journal of pharmaceutical and biomedical analysis,* 78 (2013) 14-18.
[55] Nguyen, H.P., K.A. Schug, The advantages of ESI-MS detection in conjunction with HILIC mode separations: Fundamentals and applications, *Journal of separation science,* 31 (2008) 1465-1480.
[56] Wang, N.-H., W.-L. Lee, G.-R. Her, Signal enhancement for peptide analysis in liquid chromatography–electrospray ionization mass spectrometry with trifluoroacetic acid containing mobile phase by postcolumn electrophoretic mobility control, *Analytical chemistry,* 83 (2011) 6163-6168.
[57] Contrepois, K., L. Jiang, M. Snyder, Optimized Analytical Procedures for the Untargeted Metabolomic Profiling of Human Urine and Plasma by Combining Hydrophilic Interaction (HILIC) and Reverse-Phase Liquid Chromatography (RPLC)–Mass Spectrometry*[S], *Molecular & Cellular Proteomics,* 14 (2015) 1684-1695.
[58] Gika, H.G., G.A. Theodoridis, I.D. Wilson, Hydrophilic interaction and reversed-phase ultra-performance liquid chromatography TOF-MS for metabonomic analysis of Zucker rat urine, *Journal of separation science*, 31 (2008) 1598-1608.
[59] Koch, W., S. Forcisi, R. Lehmann, P. Schmitt-Kopplin, Sensitivity improvement in hydrophilic interaction chromatography negative mode electrospray ionization mass spectrometry using 2-(2-methoxyethoxy) ethanol as a post-column modifier for non-targeted metabolomics, *Journal of Chromatography A,* 1361 (2014) 209-216.
[60] Sillner, N., A. Walker, E.-M. Harrieder, P. Schmitt-Kopplin, M. Witting, Development and application of a HILIC UHPLC-MS method for polar fecal metabolome profiling, *Journal of Chromatography B,* 1109 (2019) 142-148.
[61] Kumar, N., N. Goel, Phenolic acids: Natural versatile molecules with promising therapeutic applications, *Biotechnology Reports,* 24 (2019) e00370.
[62] Muller, M., A.G. Tredoux, A. de Villiers, Application of kinetically optimised online HILIC× RP-LC methods hyphenated to high resolution MS for the analysis of natural phenolics, *Chromatographia*, 82 (2019) 181-196.
[63] Sarrut, M., A. D'Attoma, S. Heinisch, Optimization of conditions in on-line comprehensive two-dimensional reversed phase liquid chromatography. Experimental comparison with one-dimensional reversed phase liquid chromatography for the separation of peptides, *Journal of Chromatography A,* 1421 (2015) 48-59.
[64] Sarrut, M., F. Rouvière, S. Heinisch, Theoretical and experimental comparison of one dimensional versus on-line comprehensive two dimensional liquid chromato-

graphy for optimized sub-hour separations of complex peptide samples, *Journal of Chromatography A,* 1498 (2017) 183-195.

[65] Muller, M., A.G. Tredoux, A. de Villiers, Predictive kinetic optimisation of hydrophilic interaction chromatography× reversed phase liquid chromatography separations: Experimental verification and application to phenolic analysis, *Journal of Chromatography A,* 1571 (2018) 107-120.

[66] Belitz, H.-D., W. Grosch, P. Schieberle, *Food chemistry,* Springer Science & Business Media2008.

[67] Jia, S., Y.P. Kang, J.H. Park, J. Lee, S.W. Kwon, Simultaneous determination of 23 amino acids and 7 biogenic amines in fermented food samples by liquid chromatography/quadrupole time-of-flight mass spectrometry, *Journal of Chromatography A,* 1218 (2011) 9174-9182.

[68] Gökmen, V., A. Serpen, B.A. Mogol, Rapid determination of amino acids in foods by hydrophilic interaction liquid chromatography coupled to high-resolution mass spectrometry, *Analytical and Bioanalytical Chemistry,* 403 (2012) 2915-2922.

[69] Tsochatzis, E., M. Papageorgiou, S. Kalogiannis, Validation of a HILIC UHPLC-MS/MS method for amino acid profiling in triticum species wheat flours, *Foods,* 8 (2019) 514.

[70] Zhou, G., H. Pang, Y. Tang, X. Yao, X. Mo, S. Zhu, S. Guo, D. Qian, Y. Qian, S. Su, Hydrophilic interaction ultra-performance liquid chromatography coupled with triple-quadrupole tandem mass spectrometry for highly rapid and sensitive analysis of underivatized amino acids in functional foods, *Amino Acids,* 44 (2013) 1293-1305.

[71] Tsochatzis, E.D., O. Begou, H.G. Gika, P.D. Karayannakidis, S. Kalogiannis, A hydrophilic interaction chromatography-tandem mass spectrometry method for amino acid profiling in mussels, *Journal of Chromatography B,* 1047 (2017) 197-206.

[72] Fountoulakis, M., H.-W. Lahm, Hydrolysis and amino acid composition analysis of proteins, *Journal of chromatography A,* 826 (1998) 109-134.

[73] Gika, H.G., G.A. Theodoridis, U. Vrhovsek, F. Mattivi, Quantitative profiling of polar primary metabolites using hydrophilic interaction ultrahigh performance liquid chromatography–tandem mass spectrometry, *Journal of Chromatography A,* 1259 (2012) 121-127.

[74] Choi, J.-S., A.J. Berdis, Nucleoside transporters: biological insights and therapeutic applications, *Future medicinal chemistry,* 4 (2012) 1461-1478.

[75] Indy, H.E., D.C. Woollar, Determination of orotic acid, uric acid, and creatinine in milk by liquid chromatography, *Journal of AOAC International,* 87 (2004) 116-122.

[76] Zuo, R., S. Zhou, Y. Zuo, Y. Deng, Determination of creatinine, uric and ascorbic acid in bovine milk and orange juice by hydrophilic interaction HPLC, *Food chemistry,* 182 (2015) 242-245.

[77] Broncano, J., J. Otte, M. Petrón, V. Parra, M. Timón, Isolation and identification of low molecular weight antioxidant compounds from fermented "chorizo" sausages, *Meat science,* 90 (2012) 494-501.

[78] Porter, K., J.K. Lodge, Determination of selected water-soluble vitamins (thiamine, riboflavin, nicotinamide and pyridoxine) from a food matrix using hydrophilic interaction liquid chromatography coupled with mass spectroscopy, *Journal of Chromatography B,* 1171 (2021) 122541.

[79] DeFelice, S.L. The nutraceutical revolution: its impact on food industry R&D, *Trends in Food Science & Technology*, 6 (1995) 59-61.

[80] Racine, K.C., A.H. Lee, A.C. Stewart, K.W. Blakeslee, A.P. Neilson, Development of a rapid ultra performance hydrophilic interaction liquid chromatography tandem mass spectrometry method for procyanidins with enhanced ionization efficiency, *Journal of Chromatography A,* 1594 (2019) 54-64.

[81] Yang, W., X. Ma, O. Laaksonen, W. He, H. Kallio, B. Yang, Effects of latitude and weather conditions on proanthocyanidins in blackcurrant (Ribes nigrum) of Finnish commercial cultivars, *Journal of agricultural and food chemistry,* 67 (2019) 14038-14047.

[82] Yang, W., O. Laaksonen, H. Kallio, B. Yang, Effects of latitude and weather conditions on proanthocyanidins in berries of Finnish wild and cultivated sea buckthorn (*Hippophaë rhamnoides* L. ssp. rhamnoides), *Food Chemistry,* 216 (2017) 87-96.

[83] Hollands, W.J., S. Voorspoels, G. Jacobs, K. Aaby, A. Meisland, R. Garcia-Villalba, F. Tomas-Barberan, M.K. Piskula, D. Mawson, I. Vovk, Development, validation and evaluation of an analytical method for the determination of monomeric and oligomeric procyanidins in apple extracts, *Journal of Chromatography A*, 1495 (2017) 46-56.

[84] Sommella, E., G. Pepe, F. Pagano, C. Ostacolo, G.C. Tenore, M.T. Russo, E. Novellino, M. Manfra, P. Campiglia, Detailed polyphenolic profiling of Annurca apple (M. pumila Miller cv Annurca) by a combination of RP-UHPLC and HILIC, both hyphenated to IT-TOF mass spectrometry, *Food Research International,* 76 (2015) 466-477.

[85] Küllenberg, D., L.A. Taylor, M. Schneider, U. Massing, Health effects of dietary phospholipids, *Lipids in health and disease,* 11 (2012) 1-16.

[86] Molendi-Coste, O., V. Legry, I.A. Leclercq, Why and how meet n-3 PUFA dietary recommendations? *Gastroenterology research and practice,* 2011 (2010).

[87] Parchem, K., S. Sasson, C. Ferreri, A. Bartoszek, Qualitative analysis of phospholipids and their oxidised derivatives–used techniques and examples of their applications related to lipidomic research and food analysis, *Free Radical Research,* 53 (2019) 1068-1100.

[88] Peterson, B.L., B.S. Cummings, A review of chromatographic methods for the assessment of phospholipids in biological samples, *Biomedical Chromatography,* 20 (2006) 227-243.

[89] Lísa, M., E. Cífková, M. Holčapek, Lipidomic profiling of biological tissues using off-line two-dimensional high-performance liquid chromatography–mass spectrometry, *Journal of chromatography A,* 1218 (2011) 5146-5156.

[90] Dugo, P., N. Fawzy, F. Cichello, F. Cacciola, P. Donato, L. Mondello, Stop-flow comprehensive two-dimensional liquid chromatography combined with mass

spectrometric detection for phospholipid analysis, *Journal of Chromatography A,* 1278 (2013) 46-53.

[91] Kim, J., P.E. Minkler, R.G. Salomon, V.E. Anderson, C.L. Hoppel, Cardiolipin: characterization of distinct oxidized molecular species, *Journal of lipid research,* 52 (2011) 125-135.

[92] Parchem, K., B. Kusznierewicz, T. Chmiel, P. Maciołek, A. Bartoszek, Profiling and Qualitative Assessment of Enzymatically and Thermally Oxidized Egg Yolk Phospholipids using a Two-Step High-Performance Liquid Chromatography Protocol, *Journal of the American Oil Chemists' Society*, 96 (2019) 693-706.

[93] Anesi, A., G. Guella, A fast liquid chromatography-mass spectrometry methodology for membrane lipid profiling through hydrophilic interaction liquid chromatography, *Journal of chromatography A,* 1384 (2015) 44-52.

[94] Buré, C., S. Ayciriex, E. Testet, J.-M. Schmitter, A single run LC-MS/MS method for phospholipidomics, *Analytical and bioanalytical chemistry,* 405 (2013) 203-213.

[95] Donato, P., F. Cacciola, F. Cichello, M. Russo, P. Dugo, L. Mondello, Determination of phospholipids in milk samples by means of hydrophilic interaction liquid chromatography coupled to evaporative light scattering and mass spectrometry detection, *Journal of Chromatography A,* 1218 (2011) 6476-6482.

[96] Cajka, T., O. Fiehn, Comprehensive analysis of lipids in biological systems by liquid chromatography-mass spectrometry, TrAC *Trends in Analytical Chemistry,* 61 (2014) 192-206.

[97] Schwalbe-Herrmann, M., J. Willmann, D. Leibfritz, Separation of phospholipid classes by hydrophilic interaction chromatography detected by electrospray ionization mass spectrometry, *Journal of Chromatography A,* 1217 (2010) 5179-5183.

[98] Merlo, F., J. Montagna, F. Maraschi, A. Profumo, F. Baldanti, A. Speltini, A versatile method for multiclass determination of β-lactam drugs in urine by solid-phase extraction followed by HILIC-UV, *Journal of Chromatography Open,* 2 (2022) 100048.

[99] Hameed, E.A.A., R.A.A. Salam, G.M. Hadad, Development of an optimized HPLC method for the simultaneous determination of six compounds containing β-lactam ring in human plasma and urine using experimental design methodology, *RSC advances,* 6 (2016) 9465-9474.

[100] Legrand, T., D. Vodovar, N. Tournier, N. Khoudour, A. Hulin, Simultaneous determination of eight β-lactam antibiotics, amoxicillin, cefazolin, cefepime, cefotaxime, ceftazidime, cloxacillin, oxacillin, and piperacillin, in human plasma by using ultra-high-performance liquid chromatography with ultraviolet detection, *Antimicrobial Agents and Chemotherapy,* 60 (2016) 4734-4742.

[101] W., Huang, Q. Qiu, M. Chen, J. Shi, X. Huang, Q. Kong, D. Long, Z. Chen, S. Yan, Determination of 18 antibiotics in urine using LC-QqQ-MS/MS, *Journal of Chromatography B,* 1105 (2019) 176-183.

[102] Van Toi, P., K.V. Doan, N.N.Q. Minh, P.N. Phuong, M.D. de Jong, H.R. van Doorn, T. Pouplin, Investigation of early antibiotic use in pediatric patients with acute

respiratory infections by high-performance liquid chromatography, *Biomedical Chromatography,* 34 (2020) e4699.

[103] van Amsterdam, P., A. Companjen, M. Brudny-Kloeppel, M. Golob, S. Luedtke, P. Timmerman, The European Bioanalysis Forum community's evaluation, interpretation and implementation of the European Medicines Agency guideline on Bioanalytical Method Validation, *Bioanalysis,* 5 (2013) 645-659.

[104] Food, D. *Administration, FDA guidance for industry: bioanalytical method validation, Center for Drug Evaluation and Research,* US Department of Health and Human Services, (2001).

[105] Maruthur, N.M., E. Tseng, S. Hutfless, L.M. Wilson, C. Suarez-Cuervo, Z. Berger, Y. Chu, E. Iyoha, J.B. Segal, S. Bolen, Diabetes medications as monotherapy or metformin-based combination therapy for type 2 diabetes: a systematic review and meta-analysis, *Annals of internal medicine,* 164 (2016) 740-751.

[106] M. Sgroi, F.G. Vagliasindi, S.A. Snyder, P. Roccaro, N-Nitrosodimethylamine (NDMA) and its precursors in water and wastewater: A review on formation and removal, *Chemosphere,* 191 (2018) 685-703.

[107] ICH, H.T.G. Validation of analytical procedures: text and methodology, Q2 (R1). Current Step 4 Version, Parent Guidelines on Methodology Dated November 6 1996. Incorporated in November 2005, *International Conference on Harmonisation of Technical Requirements for Registration of Pharmaceuticals for Human Use,* Geneva, Switzerland, 2005.

[108] Cubbon, S., T. Bradbury, J. Wilson, J. Thomas-Oates, Hydrophilic interaction chromatography for mass spectrometric metabonomic studies of urine, *Analytical chemistry,* 79 (2007) 8911-8918.

[109] Danaceau, J.P., E.E. Chambers, K.J. Fountain, Hydrophilic interaction chromatography (HILIC) for LC–MS/MS analysis of monoamine neurotransmitters, *Bioanalysis,* 4 (2012) 783-794.

[110] Spagou, K., H. Tsoukali, N. Raikos, H. Gika, I.D. Wilson, G. Theodoridis, Hydrophilic interaction chromatography coupled to MS for metabonomic/ metabolomic studies, *Journal of separation science,* 33 (2010) 716-727.

[111] Yu, W., H. Jin, A. Shen, L. Deng, J. Shi, X. Xue, Y. Guo, Y. Liu, X. Liang, Purification of high-purity glycyrrhizin from licorice using hydrophilic interaction solid phase extraction coupled with preparative reversed-phase liquid chromatography, *Journal of Chromatography B,* 1040 (2017) 47-52.

[112] Liu, G., B. Luan, G. Liang, L. Xing, L. Huang, C. Wang, Y. Xu, Isolation and identification of four major impurities in capreomycin sulfate, *Journal of Chromatography A,* 1571 (2018) 155-164.

[113] Reilly, J., Q. Liu, C. Howson, K. Boss, B. Shin, M. Palermo, J. Larrow, M. Qian, T. Mann, Method Development for Reproducible Flash Purification of Pharmaceuticals by UHPLC and HPLC, *Microscopy,* (2022).

[114] Rainville, P.D., G. Theodoridis, R.S. Plumb, I.D. Wilson, Advances in liquid chromatography coupled to mass spectrometry for metabolic phenotyping, *TrAC Trends in Analytical Chemistry,* 61 (2014) 181-191.

[115] Gavaghan, C.L., E. Holmes, E. Lenz, I.D. Wilson, J.K. Nicholson, An NMR-based metabonomic approach to investigate the biochemical consequences of genetic

strain differences: application to the C57BL10J and Alpk: ApfCD mouse, *FEBS letters,* 484 (2000) 169-174.

[116] Theodoridis, G.A., H.G. Gika, E.J. Want, I.D. Wilson, Liquid chromatography–mass spectrometry based global metabolite profiling: a review, *Analytica chimica acta,* 711 (2012) 7-16.

[117] Cubbon, S., C. Antonio, J. Wilson, J. Thomas-Oates, Metabolomic applications of hilic–lc–ms, *Mass spectrometry reviews,* 29 (2010) 671-684.

[118] Clarke, C.J., J.N. Haselden, Metabolic profiling as a tool for understanding mechanisms of toxicity, *Toxicologic Pathology,* 36 (2008) 140-147.

[119] Lindon, J.C., H.C. Keun, T.M. Ebbels, J.M. Pearce, E. Holmes, J.K. Nicholson, The Consortium for Metabonomic Toxicology (COMET): *Aims, activities and achievements,* (2005).

[120] Würtz, P., A.J. Kangas, P. Soininen, D.A. Lawlor, G. Davey Smith, M. Ala-Korpela, Quantitative serum nuclear magnetic resonance metabolomics in large-scale epidemiology: a primer on-omic technologies, *American journal of epidemiology,* 186 (2017) 1084-1096.

[121] Dunn, W.B., W. Lin, D. Broadhurst, P. Begley, M. Brown, E. Zelena, A.A. Vaughan, A. Halsall, N. Harding, J.D. Knowles, Molecular phenotyping of a UK population: defining the human serum metabolome, *Metabolomics,* 11 (2015) 9-26.

[122] Lewis, M.R., J.T. Pearce, K. Spagou, M. Green, A.C. Dona, A.H. Yuen, M. David, D.J. Berry, K. Chappell, V. Horneffer-van der Sluis, Development and application of ultra-performance liquid chromatography-TOF MS for precision large scale urinary metabolic phenotyping, *Analytical Chemistry,* 88 (2016) 9004-9013.

[123] Chauve, B., D. Guillarme, P. Cléon, J.L. Veuthey, Evaluation of various HILIC materials for the fast separation of polar compounds, *Journal of separation science,* 33 (2010) 752-764.

[124] Creek, D.J., A. Jankevics, R. Breitling, D.G. Watson, M.P. Barrett, K.E. Burgess, Toward global metabolomics analysis with hydrophilic interaction liquid chromatography–mass spectrometry: improved metabolite identification by retention time prediction, *Analytical chemistry,* 83 (2011) 8703-8710.

[125] Spagou, K., I.D. Wilson, P. Masson, G. Theodoridis, N. Raikos, M. Coen, E. Holmes, J.C. Lindon, R.S. Plumb, J.K. Nicholson, HILIC-UPLC-MS for exploratory urinary metabolic profiling in toxicological studies, *Analytical chemistry,* 83 (2011) 382-390.

[126] Paris, L.P., C.H. Johnson, E. Aguilar, Y. Usui, K. Cho, L.T. Hoang, D. Feitelberg, H.P. Benton, P.D. Westenskow, T. Kurihara, Global metabolomics reveals metabolic dysregulation in ischemic retinopathy, *Metabolomics,* 12 (2016) 1-10.

[127] Sun, L., J. Liu, M. Sun, L. Lin, L. Miao, Z. Ge, B. Yang, Comprehensive metabonomic analysis of heart tissue from isoproterenol-induced myocardial infarction rat based on reversed-phase and hydrophilic interaction chromatography coupled to mass spectrometry, *Journal of separation science,* 40 (2017) 2198-2206.

[128] Gray, N., K. Adesina-Georgiadis, E. Chekmeneva, R.S. Plumb, I.D. Wilson, J.K. Nicholson, Development of a rapid microbore metabolic profiling ultraperformance liquid chromatography–mass spectrometry approach for high-throughput phenotyping studies, *Analytical chemistry,* 88 (2016) 5742-5751.

[129] Coen, M., P.M. Rademacher, W. Zou, M. Scott, P.E. Ganey, R. Roth, S.D. Nelson, Comparative NMR-based metabonomic investigation of the metabolic phenotype associated with tienilic acid and tienilic acid isomer, *Chemical Research in Toxicology,* 25 (2012) 2412-2422.

[130] Tyteca, E., J.-L. Veuthey, G. Desmet, D. Guillarme, S. Fekete, Computer assisted liquid chromatographic method development for the separation of therapeutic proteins, *Analyst,* 141 (2016) 5488-5501.

[131] Le Mapihan, K., J. Vial, A. Jardy, Reversed-phase liquid chromatography column testing and classification: Physicochemical interpretation based on a wide set of stationary phases, *Journal of chromatography A,* 1144 (2007) 183-196.

[132] Neue, U.D., C.H. Phoebe, K. Tran, Y.-F. Cheng, Z. Lu, Dependence of reversed-phase retention of ionizable analytes on pH, concentration of organic solvent and silanol activity, *Journal of Chromatography A,* 925 (2001) 49-67.

[133] Kaliszan, R. QSRR: quantitative structure-(chromatographic) retention relationships, *Chemical reviews,* 107 (2007) 3212-3246.

[134] Héberger, K. Quantitative structure–(chromatographic) retention relationships, *Journal of chromatography A,* 1158 (2007) 273-305.

[135] Taraji, M., P.R. Haddad, R.I. Amos, M. Talebi, R. Szucs, J.W. Dolan, C.A. Pohl, Chemometric-assisted method development in hydrophilic interaction liquid chromatography: *A review, Analytica chimica acta, 1*000 (2018) 20-40.

[136] McCalley, D.V. Study of the selectivity, retention mechanisms and performance of alternative silica-based stationary phases for separation of ionised solutes in hydrophilic interaction chromatography, *Journal of Chromatography A,* 1217 (2010) 3408-3417.

[137] Crnogorac, G., S. Schmauder, W. Schwack, Trace analysis of dithiocarbamate fungicide residues on fruits and vegetables by hydrophilic interaction liquid chromatography/tandem mass spectrometry, Rapid Communications in Mass Spectrometry: *An International Journal Devoted to the Rapid Dissemination of Up-to-the-Minute Research in Mass Spectrometry,* 22 (2008) 2539-2546.

[138] Oellig, C., W. Schwack, Comparison of HILIC columns for residue analysis of dithiocarbamate fungicides, *Journal of Liquid Chromatography & Related Technologies,* 40 (2017) 415-418.

[139] Crnogorac, G., W. Schwack, Determination of dithiocarbamate fungicide residues by liquid chromatography/mass spectrometry and stable isotope dilution assay, Rapid Communications in Mass Spectrometry: *An International Journal Devoted to the Rapid Dissemination of Up-to-the-Minute Research in Mass Spectrometry,* 21 (2007) 4009-4016.

[140] Sanco, D. *Monitoring of pesticide residues in products of plant origin in European union, Norway, Iceland and Liechtenstein,* 2005 Report, 2007.

[141] Dayan, F., C. Cantrell, S. Duke, *Bioorg. Med. Chem,* 2009, 17 (2009) 4022.

[142] Jewell, T., D. Buffin, *Health and environmental impacts of glufosinate ammonium, Pesticide Action Network UK.* Report Prepared for Friends of the Earth UK. Pesticide Action Network UK, London, GB, (2001).

[143] Moore, J.C., J. Spink, M. Lipp, Development and application of a database of food ingredient fraud and economically motivated adulteration from 1980 to 2010, *Journal of food science,* 77 (2012) R118-R126.

[144] MacMahon, S., T.H. Begley, G.W. Diachenko, S.A. Stromgren, A liquid chromatography–tandem mass spectrometry method for the detection of economically motivated adulteration in protein-containing foods, *Journal of Chromatography A,* 1220 (2012) 101-107.

[145] Abernethy, G., K. Higgs, Rapid detection of economic adulterants in fresh milk by liquid chromatography–tandem mass spectrometry, *Journal of Chromatography A,* 1288 (2013) 10-20.

[146] Draher, J., V. Pound, T.M. Reddy, Validation of a rapid method of analysis using ultrahigh-performance liquid chromatography-tandem mass spectrometry for nitrogen-rich adulterants in nutritional food ingredients, *Journal of Chromatography A,* 1373 (2014) 106-113.

[147] Hobohm, U., T. Houthaeve, C. Sander, Amino acid analysis and protein database compositional search as a rapid and inexpensive method to identify proteins, *Analytical biochemistry,* 222 (1994) 202-209.

[148] Alpert, A.J. Hydrophilic-interaction chromatography for the separation of peptides, nucleic acids and other polar compounds, *Journal of chromatography A,* 499 (1990) 177-196.

[149] Guo, S., J.-a. Duan, D. Qian, Y. Tang, Y. Qian, D. Wu, S. Su, E. Shang, Rapid determination of amino acids in fruits of Ziziphus jujuba by hydrophilic interaction ultra-high-performance liquid chromatography coupled with triple-quadrupole mass spectrometry, *Journal of agricultural and food chemistry,* 61 (2013) 2709-2719.

[150] Prinsen, H.C., B. Schiebergen-Bronkhorst, M. Roeleveld, J. Jans, M. de Sain-van der Velden, G. Visser, P. Van Hasselt, N. Verhoeven-Duif, Rapid quantification of underivatized amino acids in plasma by hydrophilic interaction liquid chromatography (HILIC) coupled with tandem mass-spectrometry, *Journal of Inherited Metabolic Disease: Official Journal of the Society for the Study of Inborn Errors of Metabolism,* 39 (2016) 651-660.

[151] Joyce, R., V. Kuziene, X. Zou, X. Wang, F. Pullen, R.L. Loo, Development and validation of an ultra-performance liquid chromatography quadrupole time of flight mass spectrometry method for rapid quantification of free amino acids in human urine, *Amino Acids,* 48 (2016) 219-234.

[152] Krumpochova, P., B. Bruyneel, D. Molenaar, A. Koukou, M. Wuhrer, W. Niessen, M. Giera, Amino acid analysis using chromatography–mass spectrometry: An inter platform comparison study, *Journal of pharmaceutical and biomedical analysis,* 114 (2015) 398-407.

[153] Kambhampati, S., J. Li, B.S. Evans, D.K. Allen, Accurate and efficient amino acid analysis for protein quantification using hydrophilic interaction chromatography coupled tandem mass spectrometry, *Plant Methods,* 15 (2019) 1-12.

[154] Gao, J., R. Helmus, C. Cerli, B. Jansen, X. Wang, K. Kalbitz, Robust analysis of underivatized free amino acids in soil by hydrophilic interaction liquid chromatography coupled with electrospray tandem mass spectrometry, *Journal of Chromatography* A, 1449 (2016) 78-88.

[155] Allen, D.K. *Assessing compartmentalized flux in lipid metabolism with isotopes, Biochimica et Biophysica Acta (BBA)-Molecular and Cell Biology of Lipids,* 1861 (2016) 1226-1242.

[156] Allen, D.K. Quantifying plant phenotypes with isotopic labeling & metabolic flux analysis, *Current opinion in biotechnology,* 37 (2016) 45-52.

[157] Boersema, P.J., S. Mohammed, A.J. Heck, Hydrophilic interaction liquid chromatography (HILIC) in proteomics, *Analytical and bioanalytical chemistry,* 391 (2008) 151-159.

[158] Hughes, C.S., S. Moggridge, T. Müller, P.H. Sorensen, G.B. Morin, J. Krijgsveld, Single-pot, solid-phase-enhanced sample preparation for proteomics experiments, *Nature protocols,* 14 (2019) 68-85.

[159] Lindner, H., B. Sarg, W. Helliger, Application of hydrophilic-interaction liquid chromatography to the separation of phosphorylated H1 histones, *Journal of Chromatography A,* 782 (1997) 55-62.

[160] Zhang, Z., Z. Wu, M.J. Wirth, Polyacrylamide brush layer for hydrophilic interaction liquid chromatography of intact glycoproteins, *Journal of Chromatography* a, 1301 (2013) 156-161.

[161] D'Atri, V., S. Fekete, A. Beck, M. Lauber, D. Guillarme, Hydrophilic interaction chromatography hyphenated with mass spectrometry: a powerful analytical tool for the comparison of originator and biosimilar therapeutic monoclonal antibodies at the middle-up level of analysis, *Analytical chemistry*, 89 (2017) 2086-2092.

[162] Periat, A., S. Fekete, A. Cusumano, J.-L. Veuthey, A. Beck, M. Lauber, D. Guillarme, Potential of hydrophilic interaction chromatography for the analytical characterization of protein biopharmaceuticals, *Journal of Chromatography A,* 1448 (2016) 81-92.

[163] Domínguez-Vega, E., S. Tengattini, C. Peintner, J. van Angeren, C. Temporini, R. Haselberg, G. Massolini, G.W. Somsen, High-resolution glycoform profiling of intact therapeutic proteins by hydrophilic interaction chromatography-mass spectrometry, *Talanta,* 184 (2018) 375-381.

[164] Tengattini, S., E. Domínguez-Vega, C. Temporini, T. Bavaro, F. Rinaldi, L. Piubelli, L. Pollegioni, G. Massolini, G.W. Somsen, Hydrophilic interaction liquid chromatography-mass spectrometry as a new tool for the characterization of intact semi-synthetic glycoproteins, *Analytica chimica acta,* 981 (2017) 94-105.

[165] Tran, J.C., L. Zamdborg, D.R. Ahlf, J.E. Lee, A.D. Catherman, K.R. Durbin, J.D. Tipton, A. Vellaichamy, J.F. Kellie, M. Li, Mapping intact protein isoforms in discovery mode using top-down proteomics, *Nature,* 480 (2011) 254-258.

[166] Close, E.D., A.O. Nwokeoji, D. Milton, K. Cook, D.M. Hindocha, E.C. Hook, H. Wood, M.J. Dickman, Nucleic acid separations using superficially porous silica particles, *Journal of Chromatography A,* 1440 (2016) 135-144.

[167] Goyon, A., K. Zhang, Characterization of antisense oligonucleotide impurities by ion-pairing reversed-phase and anion exchange chromatography coupled to hydrophilic interaction liquid chromatography/mass spectrometry using a versatile two-dimensional liquid chromatography setup, *Analytical chemistry,* 92 (2020) 5944-5951.

Chapter 4

Supercritical Fluid Chromatography: Its Hyphenated Techniques and Applications

Indu Passi[*,1]
and Sushant Salwan[2]
[1]Department of Pharmaceutical Analysis, ISF College of Pharmacy, Ghal Kalan, Moga, Punjab, India
[2]Faculty of Pharmaceutical Sciences, The ICFAI University, Baddi, Himachal Pradesh, India

Abstract

Supercritical fluid chromatography (SFC) has had a rebirth in recent years, which has been sparked by the release of a new generation of instruments and columns from the major manufacturers of chromatographic instrumentation, who are steadfastly committed to furthering the technology. Using SFC's sophisticated features, considerable analytical work has recently been reported and is now being done. It has advantages over conventional liquid chromatographic techniques because of its low solvent consumption, excellent sensitivity, and solvent recovery. This technique can be applied to a greater variety of analytes because it makes use of supercritical fluids, which have both liquid and gaseous properties. Hyphenation with mass spectroscopy (MS), ultraviolet (UV), high-resolution mass spectrometry (HRMS), photodiode array detector (PDA), nuclear magnetic resonance (NMR), charged aerosol detector (CAD), etc. are examples of SFC techniques that have advanced. The instruments and experimental conditions to carry out "advanced SFC" are covered in the current chapter, including stationary phase, mobile phase nature, pressure, and temperature. Critical discussion is held over the use of SFC in pharmaceutical analysis,

* Corresponding Author's Email: indu.passi4@gmail.com.

In: Modern Approaches in Fluid Chromatography
Editors: Rohit Bhatia and Bhupinder Kumar
ISBN: 979-8-88697-856-8

including the identification of pharmaceuticals in formulations and various other applications.

Keywords: SFC, enantio-resolution, metabolite, polycyclic, steroids

1. Introduction

The pharmaceutical industries require highly effective analytical separation technologies since drug analysis is required at every stage of the medication development process. For this goal, a variety of chromatographic techniques are commonly used, and among them, Supercritical Fluid Chromatography (SFC) has piqued attention in recent years [1]. SFC (Supercritical fluid chromatography) has been proven to be a valuable tool for both analytical and preparative work. It has proven itself as a reliable tool for LC and GC analysis. The discovery of SFC is credited to Klesper et al. [2]. The separation of thermo-labile porphyrin derivatives using supercritical chlorofluoromethanes at pressures up to 140 bar and temperatures between 150 – 170°C was described in 1962. Theoretically and experimentally, this approach was later expanded by other researchers in the 1960s [3, 4]. It is generally agreed that Hewlett- Packard's introduction of SFC instrumentation for packed column SFC at the Pittsburgh Conference in 1981-1982, together with numerous later investigations by Berger and Gere [5, 6], marked the beginning of the SFC revival. Novotny et al. published the first report on the use of open tubular wall-coated capillary columns in SFC at the same time as this occurrence [7]. As may be seen from the growing number of papers in this area, including those on applications in preparative separation, the petroleum sector, and the assessment of physicochemical parameters, SFC has found a place among chromatographic techniques in the interim. Supercritical gases are already employed as extractants on an industrial scale in fluid extraction, which has made astounding advancements that have aided this evolution [8, 9]. Supercritical $CO_2$is the main eluent used in SFC. The critical pressure and temperature for this chemical are both within acceptable ranges (73.8 bar and ambient temperature conditions i.e., 31.1°C) [10]. SCFs are also known as "green solvents" since they are energy- and environmentally friendly. SFC was not successful in gaining popularity at that time for a number of reasons, including the fact that most research at the time was concentrated on HPLC and that GC instruments were easily accessible. SFC can be considered an older technology or even a technique than HPLC. Therefore, the chemicals that GC could not separate could be extracted using HPLC, which reduced interest in SFC [11]. This generation of SFC equipment is now referred to as Ultra-high performance Supercritical fluid chromatography (UHPSFC) systems due to the recent availability of chromatographic apparatus with various columns and cutting-edge systems [12]. Some of the shortcomings of conventional SFC equipment, such as reduced UV sensitivity, poor quantitative performance, low reliability, and resilience, were eliminated or at least much reduced by adopting cutting-edge current SFC instruments [13].

Reversed-phase chromatography typically necessitates a solvent exchange in order to elute the organic solvent or an organic extract dilution in a highly aqueous phase [14, 15]. As a result, low-polar compounds precipitate or become stuck to the sample vessel wall and the LC injector, which leads to poor method performance. Direct organic extract injection is permitted by UHPSFC, which uses supercritical CO2 with a small amount of organic solvent functioning as the mobile phase. This results in lower drug loss and lower detection limits, and along with that, it reduces analysis time by improving separation efficiency. Due to its non-reactivity and lack of toxicity, carbon dioxide has less of an adverse effect on the environment, safety, and expenses [16, 17].

Supercritical fluid chromatography is one of the technique which have potential to analyze a wide range of analytes. This technique quantifies the analytes even in ultra-small concentrations in complex matrices. The renewed interest in SFC seen in the pharmaceutical analysis community during the last couple of years is mainly driven by the introduction of new state-of-the-art systems by a number of key instrument manufacturers. The design of the newly developed devices is essentially based on recent UHPLC instruments, which already incorporated the elements which overcome several challenges that prevented SFC from being a robust technique. Various other advanced techniques in SFC include hyphenation with mass spectroscopy, high-resolution mass spectroscopy, ultraviolet spectroscopy, nuclear magnetic resonance, photodiode array detector, charged aerosol detector, etc. [18]. The present chapter highlights the working of SFC, various solvents used in SFC techniques along with the recent applications in various analytical fields.

1.1. Basic Principle of Super Critical Fluid Chromatography

Supercritical fluids (SCF) can facilitate the rotation of eluent near its critical point. The temperature at which a liquid will not remain in the liquid state no matter how much pressure is applied is known as the supercritical temperature. Additionally, there is a pressure above which a chemical cannot exist in a gaseous state, regardless of the high temperatures used. Initially, the eluent (mostly CO_2 gas) is brought to the required pressure and its solvent power and separation selectivity are adjusted (the rule $P_c < P < 4P_c$, here P_c stands for critical pressure) [19]. The basic principle of SFC can be well understood in Figure 1.

Eluent is then directed onto a chromatographic column whose temperature is kept at or near the critical temperature (Tc), where the supercritical fluid will have the greatest impact. Currently, supercritical $CO_2$is combined with co-solvent to boost its solvent power and selectivity, which also has an impact on the product's solubility. Additionally, the sample is transported from the injection port to the column by the mobile phase (CO_2 and co-solvent), where separation occurs before the analytes are identified by the detector. The solute may remain with the co-solvent and can then be separated if the co-solvent is employed; however, if the co-solvent is not used, solute precipitation occurs as soon as the supercritical fluid enters the gas phase. This can happen when the mobile phase is decompressed below its critical pressure. The subsequent step of recycling gaseous eluent involves pressurized liquefaction [19, 20].

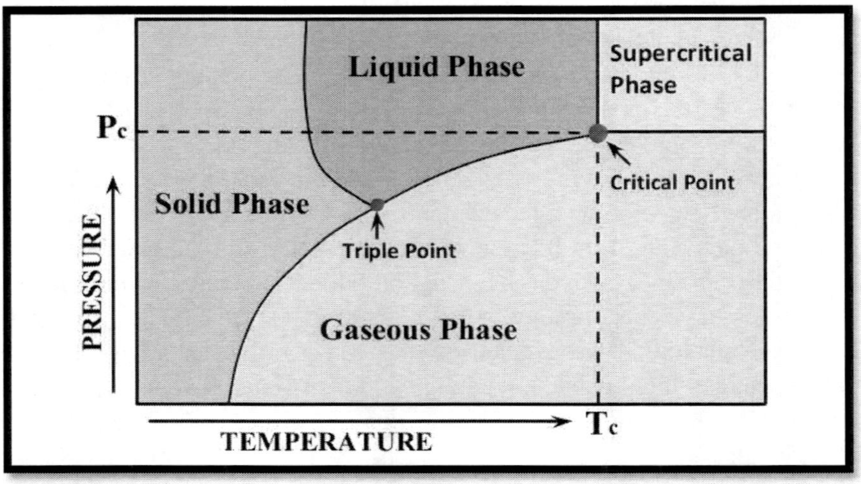

Figure 1. Basic principle of SFC; A schematic diagram describing the solid phase, liquid phase, and gas phase intercepting on the triplet point. Liquid and gas phase intercepting on the supercritical phase or critical point.

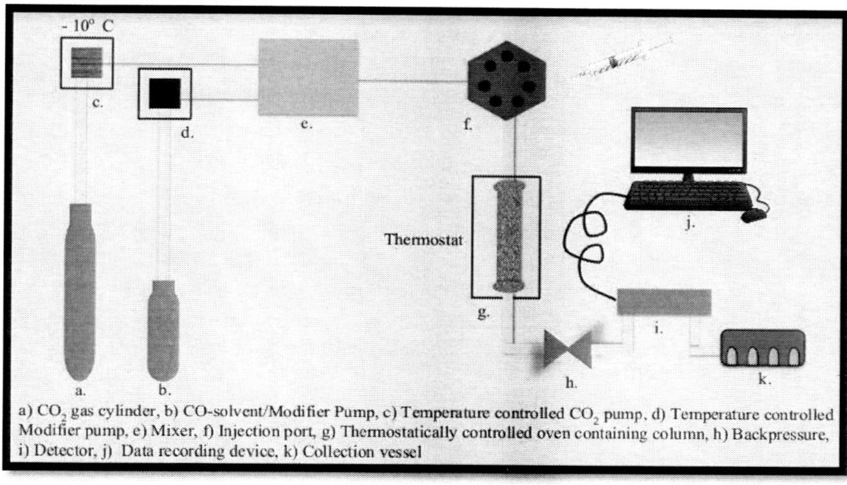

Figure 2. SFC instrumentation.

2. Instrumentation of Supercritical Fluid Chromatography

Based on the same mechanism, the movement of analytes through a stationary phase utilizing a moving mobile phase, all types of chromatography achieve the same goal that is, the separation of complicated mixtures. Due to these commonalities, commercially available HPLC systems can be configured to handle supercritical fluids with very slight alterations [21]. The physical state of the mobile phase being used liquid, gas, or vapour determines the differences between GC, HPLC, and SFC, as well as the type of column (packed or capillary), temperature control (requiring a column oven), the range of compatible detectors, and additional hardware requirements (backpressure regulation) [22].

SFC equipment might have packed columns or open tubular columns similar to GC or HPLC. The mobile phase container, a pumping system, a mixer, an injection port, a column compartment, a detector, a back pressure regulator (BPR), microprocessors, etc. are a few examples of SFC's various parts [18]. A typical SCF system instrumentation is described in (Figure 2).

Even though a number of supercritical gases have been successfully used (such as light hydrocarbons, chlorofuorocarbons, N_2O, and even NH_3), safety concerns have cemented CO_2's position as the standard for modern SFC [23]. The stainless-steel cylinder is filled with pressurized CO_2, the most widely used supercritical fluid. The pressure is kept at 55 bar at 20oC but can be varied with temperature [24]. Stainless steel cylinders are preferable. Modifiers, also known as co-solvents, are employed nowadays to improve solvent power and analyte selectivity. Most often, organic solvents including acetonitrile, isopropyl alcohol, and methanol are utilized. Since $CO_2$is insoluble in polar molecules, the use of modifiers aids in the elution of polar or ionic analytes. A pumping system pumps CO_2 fluid and modifiers in the direction of the column. The constant rate flow without any pulses that pumping systems give is a necessary condition for SFC. Typically, the type of column is taken into consideration while choosing pressure pumps. Depending on the type of column, either a capillary column or a reciprocating column, syringe pumps or reciprocating pumps are employed. There is also a secondary pump that is utilized as a modifier pump for pumping out the modifier. In order to keep CO_2 in a liquid form, the temperature of the CO_2 pump head is kept at -10°C. While the modifier is kept at 25°C to ensure a constant flow rate [25-29]. Inside the mixer, CO_2 and the co-solvent are homogenized. Mixers are similar to columns but smaller in size and are made up of tiny millimeter-sized steel balls. They also perform another crucial task, which is to balance the flow pressure irregularities caused by the pump strokes in the mixer by adjusting the mixer's volume [24].

The sample is injected in to the system through an injection valve. The sample is first dissolved in a little amount of solvent and then injected manually or automatically using a syringe. SFC offers the advantage of including nonvolatile and thermally unstable analytes, unlike GC. For many forensic and biological biomarkers, however, GC provides automatic access to volatile analytes in the gas phase above liquid samples (headspace) [30]. Different injection valves are used to inject the sample on the basis of the column being used. Pneumatic valves are used for capillary columns while injection valves similar to those used in liquid chromatography are used for packed columns. For the injection method, use a loop with a fixed injection volume or one with a variable injection volume that introduces the sample in various volume proportions [31]. The sample is transported from the injection port to the column by mobile phase as it passes through the pumps. The kind of stationary phase employed in the column affects how long analytes are retained. There are numerous stationary phase types, such as silica, alumina, polystyrene, and other highly critical CO_2 insoluble stationary phases, available with various compositions and polarity. Currently, octadecylsilyl (C-18) bonded silica columns are employed extensively. SFC columns come in two varieties: packed columns that resemble HPLC columns and capillary columns that resemble GCs. Both of these are heavily utilized in SFC. Packed columns resemble HPLC stationary phase-filled tubes. On the other hand, stationary phase is attached to fused silica capillary columns. A thermostatically controlled oven, similar to those used in GC and HPLC, is utilized to precisely manage the column's temperature in order to achieve optimal efficiency [31, 32]. Detectors are necessary for the evaluation of the analytes present in the sample. The detector components ought to be able to tolerate the high pressure produced in the SFC. Since SFC is well suited for detectors used in both HPLC and GC, many different types of detectors are used in SFC, including ultraviolet-visible spectrophotometric detectors, flame ionization detectors (FID), flame photometric detectors (FPD), refractive index detectors (RID), electron capture detectors (ECD), mass spectrometers (MS), and Fourier transform-infrared spectrometers (FT-IR). MS, FT-IR, and FID detectors are the most often utilized detectors. SFC offers a variety of uses thanks to its hyphenation with other technologies like UV, MS, etc. [18]. Some of the methods are displayed in (Figure 3).

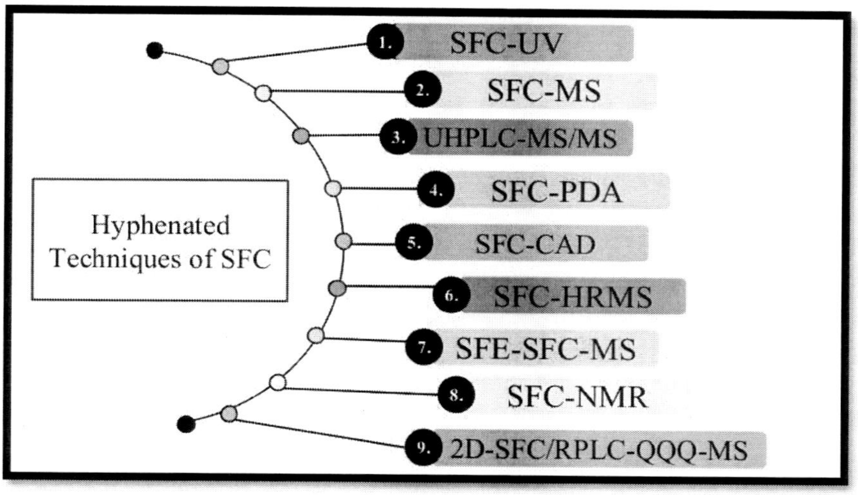

Figure 3. Hyphenated techniques of SFC.

The single-component unique to SFC is the back-pressure regulator (BPR), which can be located either in front of the detector or behind it, or both. This is because pressure and temperature both have a significant effect on the density of supercritical fluids. It is also the cause of one of SFC's biggest drawbacks, as pressure fluctuations play a significant role in explaining why

SFC is significantly less robust than UHPLC. BPRs are dual instruments with separate static and dynamic components. The dynamic half reacts to pressure variations brought on by intentional (modifier gradient, temperature gradient) and inadvertent (random) fluctuations, whereas the static part generates a constant back-pressure. Thus, dual BPRs turn the passive on-column pressure into an actively managed variable in addition to maintaining constant elution times (robustness) [33].

3. Applications of SFC in Qualitative and Quantitative Analysis

The SFC technique is used for both qualitative and quantitative analysis as well as for preparative and analytical analysis. The applications of SFC are not only limited to the pharmacy industry but are widely used in other fields too.

3.1. Pharmaceutical Industry

Enantio-resolution is a major concern for both the pharmaceutical industry and the design of drug candidates in academic laboratories. First, it is used to assess the enantiopurity of synthesis intermediates and active pharmaceutical ingredients at every stage of drug development and to purify the desired stereoisomer when the observed purity is not deemed sufficient. Secondly, it is used to assess the fate of chiral drugs, particularly isomer conversion during metabolism. It is necessary to use analytical and preparative enantio-resolution techniques. SFC is now most helpful for all of these applications because it provides a quick method [34]. Although HPLC has historically dominated the pharmaceutical industry, the benefits of SFC have long been understood, first at the preparative scale and more recently at the analytical scale. The development of more advanced analytical SFC equipment and stationary phases is largely to blame for the latter issue. Particularly, the gradual reduction in particle size of chiral stationary phases has allowed the analysis time of the screening procedure to be greatly reduced [35]. A chiral stationary phase may be helpful even if chiral separation is not necessary: Basically, a chiral stationary phase may give superior selectivity than some achiral stationary phases in any situation where structurally identical analytes are met, especially isomers (not only enantiomers). For instance, Regalado and Welch employed this technique at both the analytical and preparative scales for challenging pharmaceutical molecules [36]. Additionally, the above-mentioned recent advancements in the hyphenation to mass spectrometry are primarily responsible for the rapid rise of SFC applications in the field of bioanalysis, such as for pharmacokinetics. For instance, during the metabolization processes, the stereoisomers of an active medicinal component may interconvert. Enantio-resolution of pharmaceuticals as environmental pollutants in wastewater samples is another noteworthy illustration of the interest of SFC-MS to analyze pharmaceutical substances and their metabolites, a task that likewise requires the most sensitive detection. Another instance is the effective application of enantioselective SFC by Regalado et al. to the separation and analysis of intricate enantiomeric mixtures involving hydroxy warfarin produced by mixing warfarin with human liver microsomes [37-39].

3.2. Natural Products

Gas chromatography is typically used to resolve chiral scents, and supercritical fluid chromatography is rarely found to be competitive in this area. SFC is only useful when some transfer to a preparative scale is sought. Enantioselective SFC can be used to analyze non-volatile chiral entities found in many natural items that might be of interest in the traditional medical or food industries. The enantio-resolution of phytocannabinoids on a Whelk-O1 stationary phase served as a demonstration of the interest in sub-2 mm FPP stationary phases. The fact that this brush-type stationary phase is offered in two enantiomeric configurations, (R,R) and (S,S), enables reversed elution orders when one column is swapped out for the other [40, 41].

3.2.1. Determination of Carotenoids and Apocarotenoids in Human Blood

Most commonly, liquid extraction and liquid chromatography techniques—which require longer run times and generate a lot of organic waste—are used for the study of carotenoids and apocarotenoids. Therefore, the creation of a technique that may overcome these challenges and make it green is essential. A novel technique for quantitative profiling of the carotenoids and apocarotenoids found in both human blood and other fluids was developed by Mariosimone Zoccali et al. The developed method used a linked methodology called SFE-SFC-QqQ/MS to identify specific carotenoids and apocarotenoids in samples of human blood. These were found utilizing studies that monitored several reactions, selected ions, and available standards with full scans. Additionally, zeaxanthin, β-carotene, β-cryptoxanthin and capsanthin were analyzed directly. The average content determined of zeaxanthin was 396.9 nmol L^{-1}, for β- carotene was 123.8 nmol L^{-1}, for β-cryptoxanthin was 385.3 nmol L^{-1} and lastly for capsanthin was 38.9 nmol L^{-1} [42].

3.2.2. Quantitation of Isoflavones in Traditional Chinese Medicine

Due to their structural resemblance to estradiol, isoflavones have the ability to impact oestrogen receptors and are frequently employed for both estrogenic and antiestrogenic effects. Additionally, used as a supplement to help prevent age-related issues and diseases linked to oestrogen. They also exhibit anti-cancer, antioxidant, and anti-diabetic properties. They are phenolic phytoestrogens that come from organic sources. These are also used as standards for testing traditional Chinese medicines (TCM) that include different isomers of isoflavones. Together with his team, Wenjie Wou created a brand-new technique for inspecting the quality of TCM that contains isoflavones. He accomplished this using a UHPSFC equipped with both a PDA and an MS detector. For the measurement of isoflavones in three species of *Radix pueraria*, they continued to employ their devised approach and discovered that the highest amount was present in *Radix puerariapeduncularis* which was generally unnoticed by researchers [43].

3.2.3. Determination of Phenolic Compounds in Extra Virgin Olive Oil

Due to its abundance in monosaturated fatty acids and other nutrients including phenolic compounds, carotenoids, and tocopherols, extra virgin olive oils (EVOO) have become widely used. Additionally, they exhibit pharmacological qualities such as antibacterial, anti-inflammatory, antioxidant, and cancer and atherosclerosis prevention [44]. One-third of the phenolic components in extra virgin olive oil are phenolic acids, which may influence the oil's sensory and biological qualities. Therefore, phenolic component analysis can be used to determine the EVOO's quality [45, 46]. Out of them, numerous analytical techniques have been created. The most effective technique for phenolic compound analysis has been discovered to be HPLC. However, they have their drawbacks due to their extended run times, which are more than 60 minutes [47, 48].

3.3. Agrochemicals

About 30 percent of agrochemicals include one or more chiral centers. Even though it is recognized that the bioactivity of the stereoisomers differs greatly, racemic combinations are nevertheless used in the majority of situations. Chiral preparative techniques are required to evaluate each distinct bioactivity, whilst analytical techniques are helpful to evaluate the various rates of isomer degradation to ascertain the isomers' environmental destiny. For this reason, chiral SFC has been the subject of numerous investigations in recent years [49-51].

3.4. Applications of SFC in the Study of Lipids, Fats, and Oils

Determining triacylglycerols in milk samples is very challenging as their distribution is very complex in milk and also various isomers of triacylglycerols are present. Employment of traditional techniques only allows the identification of a limited number of major triacylglycerols. Recently, Q-TOF-MS has shown more details on triacylglycerols in milk samples. Xinghe Zhang with his group used SFC coupled with quadrupole time of flight mass spectrometry for the determination of triacylglycerols present in human milk. Through the use of Q-TOF-MS they made it possible to resolve isomers of triacylglycerols which were having the same parent ion and the same retention times. UPC ^{2} BEH-2EP column was used for the separation at 50°C. A: carbon dioxide and B: ethanol:acetonitrile (50:50,v/v) were used as mobile phases in the gradient system. They identified 95 different triacylglycerols using this method in three different samples of human milk. Although the content of triacylglycerols varies from person to person so they also tested this method on lard, peanut oil, soybean oil, walnut oil, and olive oil. In contrast to human milk, these samples contained simple triacylglycerols and were easy to separate from this method [52].

3.4.1. Lipids in HepG2 Cells

It is vital to develop lipid analysis techniques in order to examine the biological role of lipids. For the examination of several lipid classes as one peak for each of them, Hiroaki Takeda et al. employed SFC with Charged Aerosol Detector (CAD). Because ceramic phosphoethanolamine elutes at a different time than the target analytes, it was utilized as an internal standard. HepG2 cells were used to prepare the sample by lipid extraction using a modified version of Bligh and Dyer's procedure. In order to minimize background levels, DEA (Diethylamine) column was first treated to washing before being employed for analysis [53].

3.4.2. Lipids in Biological Samples

Due to their complexity in biological samples and importance as biomarkers for the diagnosis of many diseases, lipids must be analyzed using techniques that produce thorough results. With UHPSFC, liquid chromatography, and direct injection mass spectrometry, Miroslav Lisa et al. created a method for the quantification of lipid classes in various biological materials (normal tissue, malignant tissue, erythrocytes pooled plasma and plasma). Acquity BEH UPC2 column with a flow rate of 1.9 mL/min and column temperature of 60°C was the UHPSFC conditions. 1800 psi of backpressure was maintained. The modifier utilized was a gradient of methanol with water (99:1, v/v) containing 30mM ammonium acetate. There was a 6-minute runtime. For several analytes, the linearity range ranged from 1 pmol/mL to 25400 pmol/mL. With a matrix, the calibration slope ranged from 0.0156 to 14.314 and from 0.0128 to 15.1920 without a matrix. The developed UHPSFC method(analysis time-6 min) was found to be more advantageous than UHPLC (analysis time-10min) and direct injection MS method (analysis time-12 min) [54].

3.4.3. Fatty Acid Ester Toxicants (by Heat Estimation)

Industries refine edible fats and oils to remove any hazardous elements that may be present. However, these procedures result in the formation of hazardous heat-produced by-products including fatty acid esters of 3-monochloropropane-1,2-diol (3-MCPD) and 2-monochloropropane-1,3-diol (2-MCPD). According to toxicological studies on 3-MCPD metabolism, these effects include nephrotoxicity and antifertility. Therefore, the European Food Safety Agency lowered the daily consumption of MCPD from 2µg/kg of body weight to 0.8µg/kg of body weight. But there is little information available about 3-MCPD esters. In order to reduce their intake, methods of MCPD analysis in food and oil samples are required [55, 56].

3.5. Forensics

Enantio-resolution is also needed in other application areas, which is where chromatographic separation is most helpful. In the area of forensic toxicology, where the emergence of designer pharmaceuticals is necessitating the development of complementary high-performance analytical procedures like gas chromatography, HPLC, and SFC, chiral SFC has just been introduced. In fact, analyzing the impurities in pharmaceuticals—and chiral compounds in particular—can assist locate hidden labs. In a recent publication, for instance, it was demonstrated how chiral SFC techniques might be used to analyze illegal quantities of methamphetamine and its precursors [57, 58].

3.6. SFC in Drug Metabolite Analysis

3.6.1. Determination of Lumefantrine, Artemether and Its Metabolite

A fixed-dose combination of the oral medications artemethrin and lumefantrine is used to treat malaria patients. Due to the high popularity of this combination, a method to estimate the pharmacokinetic profile of these two medications was required in order to assess their safety, effectiveness, and bioavailability in biological fluids. Numerous techniques have been established in the past, however, they either had lower detection thresholds or longer analytical runs for analysing lumefantrine, artemether, and their active metabolites. A technique for the simultaneous measurement of lumefantrine, artemether, and its active metabolite dihydroartemisinin in rat plasma using SFC-MS was created by Yinxian Yang et al. They performed their work on ACQUITY UPC^{2TM} BEH-2 EP using compressed CO_2 and methanol containing 2mM ammonium acetate as the mobile phases in gradient elution and analysis was completed in only 2.5min. The linearity range for both artemether and dihydroartemisinin was 2.0-1000 ng/mL, whereas for lumefantrine it was 1.0-5000 μg/mL. Correlation coefficient for artemether, dihydroartemisinin and lumefantrine were 0.9984, 0.9971 and 0.9984 respectively. LLOQ obtained was 2.0 μg/mL for artemisinin and dihydroartemisinin and 1.0 μg/mL for lumefantrine [18, 59].

3.6.2. Determination of Flupirtine and Its Metabolites in Human Urine

Flupirtine is a muscle-relaxing analgesic that is used to treat a variety of pain conditions. Analysis of flurpirtine and its metabolites is necessary because flurpirtine's potential liver damage has been raised. For the simultaneous measurement of flupirtine and metabolites under acidic and basic conditions from human urine, Robert K. Hofstettera et al. established a quick, eco-friendly, and adaptable technique employing SFC-MS. Using the EMA's recommendations for validating bioanalytical analytical methods, they validated their approach. Retention times were 4.85, 5.03, and 3.71 for flupirtine, D-13223 (a basic metabolite), and 4-fluorohippuric acid, respectively, with r^2 values of 0.999, 0.998, and 0.997. (Acidic metabolite) [60, 61]. LLOQ was determined to be within the acceptable range as per the EMA's guidelines for validating bioanalytical methods. Targeted metabolites may be analyzed in urine using the devised approach, which was used to analyze human urine samples for pilot pharmacokinetic research.

3.6.3. Ketamine Metabolite Stereoselective Analysis in Human Urine

Ketamine has been used as an antidepressant at subanesthetic levels; its action is quick and lasts for about a week even after just one dose. Vertigo, headaches, blurred vision, and other side effects are also brought on by it [62]. The R and S enantiomers of ketamine are both chemically present and function in the same manner. But compared to R-ketamine, S-ketamine has a threefold higher affinity for NMDARs. After metabolism, ketamine produces several metabolites with different pharmacological properties, such as R- and S-norketamine, R- and S- dehydronorketamine, and 2S 6S- and 2R 6R-hydroxynorketamine [63, 64]. Therefore, an analytical technique is required to stereo-selectively measure them.

3.7. Soil Samples

The presence of various polycyclic aromatic hydrocarbons (PHAs) in all main environmental compartments, such as soil, air, and water, as well as the threat they provide to human and animal habitats, are environmental concerns today [65, 66]. PHAs are continuously condensed on particles in the air, water, and soil and are created by the incomplete combustion of organic matter and fossil fuels. The study of these chemicals' cycle behaviour in and through natural and manmade sources has been motivated by their carcinogenic and mutagenic effects. Soxhlet extraction, matrix solid-phase dispersion, microwave-assisted extraction, ultrasonic and mechanical agitation, the QuEChERS (Quick, Easy, Cheap, Effective, Rugged and Safe) method, and pressurized and supercritical fluid extraction have all been used to extract and pre-concentrate aromatic hydrocarbon compounds from soil samples [67, 68]. Although GC-MS has historically been utilized for PAH analysis, SFC-MS has also been employed recently to separate them. Wicker and colleagues have for the first time suggested an improved extraction and identification by SFE-SFC-MS in order to eliminate laborious sample preparation. They researched crucial online extraction and analysis factors, such as modifier type and concentration, static and dynamic extraction times, and extraction vessel temperature, in order to optimize the procedure. 16 standard PHAs of environmental significance were discovered and quantified on three distinct certified reference materials (CRM), including sand, clay, and sediment, to verify the chromatographic approach [68].

3.8. Isolation and Purification Using SFC

3.8.1. Purification of API and Dissociation from Degradation Products

Angeline Noireau et al. analyzed a developed API at stage 1 which has a potential for future drug development. They subjected the product to oxidative deterioration and cleaned the oxidized chemical before analyzing its activity. API was isolated from the three degradation products (two isomeric mono-oxidized and one di-oxidized). SFC condition is essential for the separation with 70% co-solvent in CO_2, which is reportedly compatible with enhanced-fluidity liquid chromatography, due to the analytes' high molecular weight, which results in limited solubility and stationary phase. The design of experiments along with the quick and thorough screening of stationary phases add an analytically optimized approach right away. Then, the optimized procedure was scaled up for preparation. Then they gather four fractions, including the API and the degradation products of two mono-oxidized and one di-oxidized molecules from 5.47 to 6.67 minutes, 6.83 to 8.3 minutes, and 11.91 to 15.43 minutes, respectively, for purification with less CO_2 consumption because it was recycled from the co-solvent. 98.5 percent of pure API was obtained after fractions analysis, although up to 2.6 percent of mono- and di-oxidized compounds were also found. They concluded that preparative SFC was excellent for removing degradation products to test potential future activity [69].

3.8.2. Protein and Peptide Purification

The use of SFC for bigger macromolecules, peptides, and proteins has not been very common. However, as the use of pharmacological therapies increased, peptides and proteins with bigger structures became more vital to their effectiveness. Nicole M. et al. investigated global conformational changes using size exclusion chromatography linked to hydrogen-deuterium exchange (SEC-HDX) and circular dichroism (CD) spectroscopy, followed by preparative SFC purification. They discovered that bradykinin and insulin can also be employed for SFC purification. By using three different orthogonal techniques, including 1) calculated percent alpha-helicity based on CD spectra, 2) alpha-helix-temperature hysteresis analysis by CD, and 3) SEC-HDX-MS at various temperatures, insulin was also able to regain its original structure in comparison to pre-purification insulin. They did a demonstration, however, that the high-order structures of three model proteins (cytochrome C, apomoglobin, and ubiquitin) were altered after SFC purification and did not restore their original forms. They added that their approach had been effectively applied to a number of medicinal peptide initiatives at their company as well as to tiny proteins [70].

3.8.3. UHPSFC-Based Synthetic Cathinone Separation

Cathinones are naturally occurring beta ketone amphetamine mimics found in the leaves of the Catha edulis plant. To create synthetic cathinones, these naturally occurring cathinones are derivatized. In several locations, synthetic cathinones are offered for sale as bath salts under the category of prohibited substances. For instance, the United States has jurisdiction over about 15 cathinones [71-73] -]. UHPSFC, UHPLC, HILIC, and GC were used in a comparative study by Stephanie Carnes et al. to separate synthetic cathinones. For the analysis of cathinones and their isomers, they collected samples from nine different bath salts. Cathinones of approximately 35 distinct positional isomers were also isolated overall. All methods were fine-tuned to yield the best outcomes before being compared. Torus Diol column with 3% methanol as a modifier and 10 mM ammonium format as additive in isocratic mode were the final UHPSFC settings after optimization. Only 11 of the 15 cathinones in the sample could be separated by UHPLC; 12 could be separated by UHPLC; 13 could be separated by GC; and only 5 of the 15 cathinones could be resolved by HILIC. However, run time, which was nearly two times shorter than that of the competition, was the key advantage of UHPSFC over GC and UHPLC. For the positional isomers, UHPSFC produced the best resolution, separating 28 out of 34; this was followed by HILIC, which offered isomer separation for 27 out of 34; after that, GC separated 22 out of 34; and lastly, RP-HPLC separated 7 out of 34. The lowest LOD ranged from 1.2 to 3.6 ng/ml with UHPSFC as the detector in all techniques, while UHPLC and GC provided LOD in ranges of 26 to 129 μg/ml and 1290 to 5770 μg/ml, respectively. In light of this, it can be said that altogether, UHPSFC offers the greatest results when compared to UHPLC and GC [74].

3.9. Miscellaneous Applications

3.9.1. Polycyclic Aromatic Compound Analysis

Unconventional oil feeds contain oxygenated polycyclic aromatic compounds (OPAC), which, if not removed from fuels during the refining process, have a deleterious impact on fuel quality. Therefore, it was necessary to create a technique for analyzing OPAC found in unusual oils. SFC was utilized by Josephine S. Lübeck et al. to separate and analyze the oxygenated polycyclic aromatic compounds found in unconventional oils used as fuels from the polycyclic aromatic hydrocarbons. Elution of the analytes, such as furans, aldehydes, ketones, phenols, and carboxylic acids, occurred in accordance with their polarity using the SFC combined with UV and Q-TOF-MS, or SFC-UV-ESI—QTOF-MS. The best separation was achieved using an ethylene bridge silica column. With mobile phase CO_2: Ethanol (0.1% formic acid) in gradient flow, a Torus 2-picolylamine column was employed [75].

When meals are heated, polycyclic aromatic hydrocarbons (PAHs) are formed in them. They are produced during the combustion of organic materials, such as during the roasting and smoking processes used in the production of food, and are naturally carcinogenic. The European Union has set limits on their quantity. Every food product that is heated must be tested for the presence of 16 distinct PAHs, according to European Union regulations. Because dark beer and coffee beverages are roasted during production, PAHs monitoring is necessary for them as well. With the help of his team, Yoshioka created a brand-new technique for analyzing polycyclic aromatic hydrocarbons found in dark beer and coffee utilizing SFC/APCI-MS and changing backpressure. The Inertsil ODS-P column was completed, and the mobile phase was CO_2: 0.1% formic acid in acetonitrile. Only 14 minutes was run time in all samples. Sample concentrations for linearity ranged from 0.5, 1, 2.5, 5, and 10 μg/kg, with correlation coefficients for various analytes ranging from 0.991 to 0.999. LOQ ranged from 0.4 to 1 μg/kg. For coffee samples, RSD ranged from 6 to 14 percent, while for beer samples, it ranged from 5 to 15%. They claimed that variations in the sample amount that arrived at the MS detector from SFC were the cause of the increased RSD. They were able to improve the signal-to-noise ratio of all 16 compounds by 2.4 to 8.0 times by controlling the back pressure [76-77].

3.9.2. Polar Compound Characterization by SFC-NMR

A significant portion of molecules, particularly in biochemistry, is composed of polar chemicals. They can be found in biological fluids as well. As a biomarker, metabolites produced inside the body can be very helpful in identifying diseases. Patients using NANS medication had higher concentrations of the polar metabolite ManNac. SFC-NMR was used by FHM Van Zelst to create a technique for analyzing polar materials. We used a synthetic urine sample to account for any discrepancies. 30-percent methanol in CO_2 and a 2-PIC column at 60°C was selected as the best chromatographic conditions. 1.5 mL/min of flow was produced at 120 bar of backpressure. Under these circumstances, it was demonstrated that this approach could distinguish between the N-acetyl-glucosamine (Glc-Ac) and N-acetyl-mannosamine isomers of N-acetylhexosamine (ManNAc). However, for SFC-NMR to be able to identify ManNac, a large concentration is necessary. It cannot, therefore, be utilized for patient samples. Modifications can be made, though, to make it possible [78].

3.9.3. Steroids Analysis

A technology that can be utilized to identify steroids is UPSFC-MS/MS. To obtain relevant and more precise data, there is a constant need for improvement in the analytical methods of steroid profiling. In order to analyze endogenous steroids of the four main groups (androgens, estrogens, progestogens, and corticosteroids) using the UPSFC-MS/MS technology, Neil de Kock et al. The calibration range for the linear regression coefficient for 19 steroid samples was 0.05-500 μg/mL in human plasma, and it was in the range 0.9983-0.9999. The intra- and inter-day findings had an RSD of less than 15%, according to the precision results. The 19 analytes' accuracy ranged from 80 to 116 percent. LOQ values ranged between 0.05 and 1.0 μg/mL. A total number of 19 steroids were detected in 5 minutes. Therefore, this method is suitable for routine analysis [79].

Conclusion

SFC looks to be a promising method for analyzing medicines in formulations or biological fluids. Advanced SFC technology offers the following advantages over RPLC, which is still regarded as the gold standard in the pharmaceutical industry: Comparable to or even superior to state-of-the-art RPLC using columns packed with sub-2 m porous or sub-3 m superficially porous particles, SFC offers kinetic performance (throughput and/or plate count). It just came to light that it has enormous potential in several analytical domains. The fundamental ideas and elements have been explored in the chapter along with a comparison to conventional methods. In addition, this study provides describes applications of SFC and its hyphenated approaches in the analysis of medicines, pesticides, metabolites, vitamins, steroids, hydrocarbons, etc. together with presented data.

References

[1] Desfontaine, V., Guillarme, D., Francotte, E., and Nováková, L. (2015). Supercritical fluid chromatography in pharmaceutical analysis, *Journal of Pharmaceutical and Biomedical Analysis*, 113, 56-71.

[2] Klesper, E., Corwin, A. H., and Turner, D. A. (1962). High pressure gas chromatography above critical temperatures, *Journal of Organic Chemistry*, 27, 700–701.

[3] Giddings, J. C., Myers, M. N., and King, J. W. (1969). Dense gas chromatography at pressures to 2000 atmospheres, *Journal of Chromatographic Science*, 7, 276.

[4] Sie, S. T., and Rijnders, G. W. A. (1967). High-pressure gas chromatography and chromatography with supercritical fluids. II. Permeability and efficiency of packed columns with high-pressure gases as mobile phases under conditions of incipient turbulence, *Separation Science*, 699.

[5] Berger, T. A. (1995). Packed Column SFC, *Royal Society of Chemistry*, Letchworth, UK.

[6] Gere D. R., Houck, R. K., Pacholec, F., and Rosselli, A. C. P. (1988). Supercritical fluid chromatography: a technology update, Fresenius, *Journal of Analytical Chemistry*, 222–224.

[7] Novotny, M., Springston, S. R., Peaden, P. A., Fjeldsted, J. C., and Lee, M. L. (1981). Capillary supercritical fluid chromatography, *Analytical Chemistry*, 407A.

[8] U VAN Wasen, Swaid, & Schneider, (1980). Physicochemical principles and applications of supercritical fluid chromatography, New analytical methods, *Angewandte Chemie International Edition in English*, 19(8), 575-587.

[9] Taylor, L. T. (2009). Supercritical fluid chromatography for the 21st century, *The Journal of Supercritical Fluids*, 47(3), 566-573.

[10] Majewski, Valery, E., and Ludemann-Hombourger. (2005). Principle and applications of supercritical fluid chromatography, *Journal of liquid chromatography & related technologies*, 1233-1252.

[11] Saito, M. (2013). History of supercritical fluid chromatography Instrumental development, *J. Biosci. Bioeng.*, 115 590–599. https://doi.org/10.1016/j.jbiosc. 2012.12.008.

[12] Nováková, L., Perrenoud, A. G., Francois, I., West, C., Lesellier, E., and Guillarme, D. (2014). Modern analytical supercritical fluid chromatography using columns packed with sub-2 μm particles: A tutorial, *Anal. Chim. Acta.*, 824, 18–35. https://doi.org/10.1016/j.aca.2014.03.034.

[13] Grand-Guillaume Perrenoud, A., Veuthey, J. L., and Guillarme, D. (2014). The use of columns packed with sub-2 μm particles in supercritical fluid chromatography, TrAC - *Trends Anal. Chem.*, 63 44–54. https://doi.org/10.1016/j.trac.2014.06.023.

[14] Baker, D. R., and Kasprzyk-Hordern, B. (2011). Critical evaluation of methodology commonly used in sample collection, storage and preparation for the analysis of pharmaceuticals and illicit drugs in surface water and wastewater by solid phase extraction and liquid chromatography-mass spectrometry, *J. Chromatogr. A.*, 8036–8059. https://doi.org/10.1016/j.chroma.2011.09.012.

[15] Pal, R., Megharaj, M., Kirkbride, K. P., and Naidu, R. (2013). Illicit drugs and the environment, *A review*, 1079–1092. https://doi.org/10.1016/j.scitotenv.2012.05.086.

[16] Kasprzyk-hordern, B., Thomas, K. V., and Camacho-mu, D. (2016). Enantioselective simultaneous analysis of selected pharmaceuticals in environmental samples by ultrahigh performance supercritical fl uid based chromatography tandem mass spectrometry, *Anal. Chim. Acta.*, 934 239–251. https://doi.org/10.1016/j.aca.2016.05.051.

[17] González-mariño, I., Thomas, K. V., and Reid, M. J. (2018). Determination of cannabinoid and synthetic cannabinoid metabolites in wastewater by liquid-liquid extraction and ultrahigh performance supercritical fluid chromatography-tandem mass spectrometry, *Drug Test. Anal.*, 10, 222–228. https://doi.org/10.1002/dta.2199.

[18] Perrut, M. (1994). Advances in supercritical fluid chromatographic processes, *J. Chromatogr.*, A. 658 293–313. https://doi.org/10.1016/0021-9673(94)80022-7.

[19] Majewski, W., Valery, E., and Ludemann-Hombourger, O. (2005). Principle and applications of supercritical fluid chromatography, *J. Liq. Chromatogr. Relat. Technol.*, 28, 1233–1252. https://doi.org/10.1081/JLC-200053039.21.

[20] Simpson, R. C., Gant, J. R., and Brown, P. R. (1986). Modifcation of conventional high-performance liquid chromatography equpment for use in carbon dioxide supercritical fuid chromatography. *J Chromatogr A*, 109–119. https://doi.org/10.1016/S0021 -9673(01)94698-9.

[21] Berger, T. A. (2017). Evolution of instrumentation for analytical scale supercritical fuid chromatography. In: *Supercritical fuid chromatography*. Elsevier, pp. 173–21.

[22] Nováková, L., Grand-Guillaume Perrenoud, A., Francois, I. et al. (2014). Modern analytical supercritical fuid chromatography using columns packed with sub-2 μm particles: a tutorial. *Anal Chim Acta*, 824, 18–35. https://doi.org/10.1016/j.aca.2014.03.034.

[23] Berger, T. A. (2015). Instrumentation for Analytical scale supercritical fluid chromatography, *J. Chromatogr. A.*, 1421, 171–83. https://doi.org/10.1007/978-94-011-2146-0_2.

[24] Turner, C., King, J. W., and Mathiasson, L. (2001). Supercritical fluid extraction and chromatography for fat-soluble vitamin analysis, *J. Chromatogr. A.*, 936, 215–237. https://doi.org/10.1016/S0021-9673(01)01082-2.

[25] Hsieh, Y., Li, F., and Duncan, C. J. G. (2007). Supercritical fluid chromatography and high-performance liquid chromatography/tandem mass spectrometric methods

for the determination of cytarabine in mouse plasma, *Anal. Chem.*, 79, 3856–3861. https://doi.org/10.1021/ac062441s.

[26] Takahashi, K., Takahashi, R., Horikawa, Y., and Matsuyama, S. (2013). Optimization of experimental parameters for separation of nonionic surfactants by supercritical fluid chromatography, *J. Supercrit. Fluids.*, 82, 256–262. https://doi.org/10.1016/j.supflu.2013.08.007.

[27] Zhang, X., Scalf, M., Westphall, M. S., and Smith, L. M. (2008). Membrane Protein Separation and Analysis by Supercritical Fluid Chromatography - Mass Spectrometry, *Anal. Chem.*, 80, 2590–2598.

[28] Greibrokk, T. (1993). Instrumentation for supercritical fluid chromatography, in: *Appl. Supercrit. Fluids Ind. Anal.*, Chapman & Hall, pp. 12–45.

[29] Mochalski, P., and Unterkofer, K. (2016). *Quantifcation of selected volatile organic compounds in human urine by gas chromatography selective reagent ionization time of fight mass spectrometry (GC-SRI-TOF-MS) coupled with head-space solid-phase microextraction (HS-SPME)*, https:// doi.org/10.1039/C6AN00825A.

[30] Sethi, N., Anand, A., Jain, G., Srinivas, K. S., and Chandrul, K. K. (2010). Supercritical Fluid Chromatography-A Hybrid of GC and LC, *Chronicles Young Sci.*, 1, 12–22. www.opubs.com/cys.

[31] Palmieri, M. D. (1988). An introduction to supercritical fluid chromatography Part 1: Principles and instrumentation, *J. Chem. Educ.*, 65 A254. https://doi.org/10.1021/ed065pa254.

[32] Guillarme, D., Desfontaine, V., Heinisch, S., and Veuthey, J. L. (2018). What are the current solutions for interfacing supercritical fluid chromatography and mass spectrometry, *J Chromatogr B*, 1083, 160–170. https://doi.org/10.1016/ j.jchromb.2018.03.010.

[33] Hicks, M. B., Regalado, E. L., Tan, F., Gong, X., and Welch, C. J. (2016). Supercritical fluid chromatography for GMP analysis in support of pharmaceutical development and manufacturing activities, *J. Pharm. Biomed. Anal.*, https://doi.org/10.1016/j.jpba.2015.09.014.

[34] Mattrey, F. T., Makarov, A. A., Regalado, E. L., Bernardoni, F., Figus, M., Hicks, M. B., Zheng, J., Wang, L., Schafer, W., Antonucci, V., Hamilton, S. E., Zawatzky, K., and Welch, C. J. (2017). Current challenges and future prospects in chromatographic method development for pharmaceutical research, TrAC *Trends Anal. Chem.*, 95, 36e46. https://doi.org/10.1016/j.trac.2017.07.021.

[35] Regalado, E. L., and Welch, C. J. (2015). Separation of achiral analytes using supercritical fluid chromatography with chiral stationary phases, TrAC *Trends Anal. Chem.*, 67 74e81. https://doi.org/10.1016/j.trac.2015.01.004.

[36] West, C. (2018). Current trends in supercritical fluid chromatography, *Anal. Bioanal. Chem.*, 410, 6441e6457. https://doi.org/10.1007/s00216-018-1267-4.

[37] Camacho-Munoz, D., Kasprzyk-Hordern, B., and Thomas, K. V. (2016). Enantioselective ~ simultaneous analysis of selected pharmaceuticals in environmental samples by ultrahigh performance supercritical fluid based chromatography tandem mass spectrometry, *Anal. Chim. Acta*, 934, 239e251. https://doi.org/ 10.1016/j.aca.2016.05.051.

[38] Regalado, E. L., Sherer, E. C., Green, M. D., Hendersonl, D. W., Thomas Williamson, R., Joyce, L. A., and Welch, C. J. (2014). Chromatographic separation and assignment of absolute configuration of hydroxywarfarin isomers: absolute configuration of hydroxywarfarins, *Chirality*, 26, 95e101. https://doi.org/10.1002/chir.22274.

[39] Mazzoccanti, G., Ismail, O. H., D'Acquarica, I., Villani, C., Manzo, C., Wilcox, M., Cavazzini, A., and Gasparrini, F. (2017). Cannabis through the looking glass: chemo- and enantio-selective separation of phytocannabinoids by enantioselective ultra high performance supercritical fluid chromatography, *Chem. Commun.*, 53 12262e12265. https://doi.org/10.1039/C7CC06999E.

[40] Raimbault, A., Dorebska, M., and West, C. (2019). A chiral unified chromatographyemass spectrometry method to analyze free amino acids, *Anal. Bioanal. Chem.*, https://doi.org/10.1007/s00216-019-01783-5.

[41] Zoccali, M., Giuffrida, D., Sala, F., Giofr, S. V., and Mondello, L. (2018). Carotenoids and apocarotenoids determination in intact human blood samples by online supercritical fluid extraction-supercritical fluid chromatography-tandem mass spectrometry, *Anal. Chim. Acta.*, 40–47. https://doi.org/10.1016/j.aca.2018.06.022.

[42] Wu, W., Zhang, Y., Zhang, F., Liu, J., Ren, Z., Xu, Y., Liu, T., Zhou, W., Li, H., and Zhang, C. (2019). An analytical strategy for accurate, rapid and sensitive quantitative analysis of isoflavones in traditional Chinese medicines using ultra-high performance supercritical fluid chromatography: take Radix Puerariae as an example, *J. Chromatogr. A.*, https://doi.org/10.1016/j.chroma.2019.460385.

[43] De A, L. (2014). *Virgin Olive Oil: Production, Composition, Uses and Benefitsfor Man*, UK edition, Nova Science Publishers Inc.

[44] Bahiyah, N., Bakar, A., Makahleh, A., and Saad, B. (2012). In-vial liquid – liquid microextraction-capillary electrophoresis method for the determination of phenolic acids in vegetable oils, *Anal. Chim. Acta.*, 59–66. https://doi.org/10.1016/j.aca.2012.02.045.

[45] Alkan, D., Tokatli, F., and Ozen, B. (2012). Phenolic Characterization and Geographical Classification of Commercial Extra Virgin Olive Oils Produced in Turkey, *J. Am. Oils Chem.*, 261–268. https://doi.org/10.1007/s11746-011-1917-6.

[46] Kesen, S., Kelebek, H., and Selli, S. (2014). LC – ESI – MS Characterization of Phenolic Profiles Turkish Olive Oils as Influenced by Geographic Origin and Harvest Year, *J. Am. Oils Chem*, 385–394. https://doi.org/10.1007/s11746-013-2380-3.

[47] Yang, F., Tang, G., Liu, S., Fan, Z., Wang, Y., Deng, H., Bian, Z., and Li, Z. (2019). *Ultra-performance convergence chromatography with tandem mass spectrometry-assisted method for rapid enantioseparation and determination of fluazifop-butyl in tobacco and soil*, https://doi.org/10.1002/ chir.23062.

[48] Tao, Y., Zheng, Z., Yu, Y., Xu, J., Liu, X., Wu, X., Dong, F., and Zheng, Y. (2018). Supercritical fluid chromatographyetandem mass spectrometry-assisted methodology for rapid enantiomeric analysis of fenbuconazole and its chiral metabolites in fruits vegetables, cereals, and soil, *Food Chem.*, https://doi.org/10.1016/j.foodchem.2017.08.038.

[49] Jiang, Z., Cao, X., Li, H., Zhang, C., Abd El-Aty, A. M., Jeong, J. H., Shao, Y., Shao, H., Jin, M., Jin, F., and Wang, J. (2018). Rapid analysis of tristyrylphenol ethoxylates in cucumber-field system using supercritical fluid chromatographyetandem mass spectrometry, *Food Chem.*, https://doi.org/ 10.1016/ j.foodchem.2018.05.122.

[50] Zhang, X., Zhao, Y., Cui, X., Wang, X., Shen, H., Chen, Z., Huang, C., Meruva, N., Zhou, L., Wang, F., Wu, L., and Luo, F. (2018). Application and enantiomeric residue determination of diniconazole in tea and grape and apple by supercritical fluid chromatography coupled with quadrupole-time-of-flight mass spectrometry, *J. Chromatogr. A*, https://doi.org/10.1016/ j.chroma.2018.10.051.

[51] Zhang, X., Qi, C., Zhang, Y., Wei, W., Jin, Q., and Xu, Z. (2018). Identification and quantification of triacylglycerols in human milk fat using ultra-performance convergence chromatography and quadrupole time-of-flight mass spectrometery with supercritical carbon dioxide as a mobile phase, *Food Chem.*, 712–720. https://doi.org/10.1016/j.foodchem.2018.09.150.

[52] Takeda, H., Takahashi, M., Hara, T., Izumi, Y., and Bamba, T. (2019). *Improved quantitation of lipid classes using supercritical fluid chromatography with a charged aerosol detector*, 1465–1474. https://doi.org/10.1194/jlr.D094516.

[53] Lísa, M., Cífková, E., Khalikova, M., Ovčačíková, M., and Holčapek, M. (2017). Lipidomic analysis of biological samples: Comparison of liquid chromatography, supercritical fluid chromatography and direct infusion mass spectrometry methods, *J. Chromatogr. A.*, 96–108. https://doi.org/10.1016/j.chroma.2017.10.022.

[54] Pudel, F., Benecke, P., Fehling, P., Freudenstein, A., Mattha, B., and Schwaf, A. (2011). On the necessity of edible oil refining and possible sources of 3-MCPD and glycidyl esters, *Eur. J. Lipid Sci. Technol*, 368–373. https://doi.org/10.1002/ejlt. 201000 460.

[55] Liu, M., Gao, B. Y., Qin, F., Wu, P. P., Shi, H. M., Luo, W., Ma, A. N., Jiang, Y. R., Xu, X. B., and Yu, L. L. (2012). Acute oral toxicity of 3-MCPD mono- and di-palmitic esters in Swiss mice and their cytotoxicity in NRK-52E rat kidney cells, *Food Chem. Toxicol*, 3785–3791. https://doi.org/10.1016/j.fct.2012.07.038.

[56] Carnes, S., O'Brien, S., Szewczak, A., Tremeau-Cayel, L., Rowe, W. F., McCord, B., and Lurie, I. S. (2017). Comparison of ultra high performance supercritical fluid chromatography, ultra high performance liquid chromatography, and gas chromatography for the separation of synthetic cathinones, *J. Sep. Sci.*, https://doi.org/10.1002/jssc.201700349.

[57] Segawa, H., Iwata, Y. T., Yamamuro, T., Kuwayama, K., Tsujikawa, K., Kanamori, T., and Inoue, H. (2019). Simultaneous chiral impurity analysis of methamphetamine and its precursors by supercritical fluid chromatographyetandem mass spectrometry, *Forensic Toxicol*, https://doi.org/10.1007/s11419- 018-0446-5.

[58] Yang, Y., Gao, H., Hou, S., Su, R., Liu, H., and Sun, J. (2018). A sensitive, high-throughput, and ecofriendly method for the determination of lumefantrine, artemether, and its active metabolite dihydroartemisinin by supercritical fluid chromatography and tandem mass spectrometry, *J. Sep. Sci*, 2688–2696. https://doi.org/10.1002/jssc.201800025.

[59] F. H.A., F. A, Flupirtine. (1993). A review of its pharmacological properties, and therapeutic efficacy in pain states, *Drugs*, 548–569.

[60] Anderson, N., and Borlak, J. (2011). *Correlation versus causation? Pharmacovigilance of the analgesic flupirtine exemplifies the need for refined spontaneous adr reporting*, https://doi.org/10.1371/journal.pone.0025221.

[61] Murrough, J. W., Iosifescu, D. V., Chang, L. C., Al Jurdi, R. K., Green, C. E., Perez, A. M., Iqbal, S., Pillemer, S., Foulkes, A., Shah, A. and Charney, D. S. (2013). Antidepressant Efficacy of Ketamine in Treatment-Resistant Major Depression: A Two-Site Randomized Controlled Trial, *Am. J. Psychiatry*, 1134–1142. https://doi.org/10.1176/appi.ajp.2013.13030392.

[62] Yang, C., Qu, Y., Abe, M., Nozawa, D., Chaki, S., and Hashimoto, K. (2017). (R)-*Ketamine Shows Greater Potency and Longer Lasting Antidepressant Effects Than Its Metabolite (2R,6R)-Hydroxynorketamine*, https://doi.org/10.1016/ j.biopsych. 2016.12.020.

[63] Zanos, P., Moaddel, R., Morris, P. J., Georgiou, P., Fischell, J., Elmer, G. I., Alkondon, M., Yuan, P., Pribut, H. J., Singh, N. S., Dossou, K. S. S., Fang, Y., Huang, X. P., Mayo, C. L., Wainer, I. W., Albuquerque, E. X., Thompson, S. M., Thomas, C. J., Zarate, C. A., and Gould, T. D. (2016). *NMDAR inhibition-independent antidepressant actions of ketamine metabolites*, 481–486. https://doi.org/10.1038/nature17998.

[64] Raza, N., Hashemi, B., Kim, K. H., Lee, S. H., and Deep, A. (2018). Aromatic hydrocarbons in air, water, and soil: Sampling and pretreatment techniques. *TrAC Trends Anal. Chem.*, 103, 56–73.

[65] Wicker, A. P., Carlton, D. D., Tanaka, K., Nishimura, M., Chen, V., Ogura, T., Hedgepeth, W., and Schug, K. A. (2018). On-line supercritical fluid extraction—supercritical fluid chromatography-mass spectrometry of polycyclic aromatic hydrocarbons in soil. *J. Chromatogr. B*, 1086, 82–88.

[66] Dimpe, K. M., and Nomngongo, P. N. (2016). Current sample preparation methodologies for analysis of emerging pollutants in different environmental matrices. *TrAC Trends Anal. Chem.*, 82, 199–207.

[67] Yoshioka, T., Nagatomi, Y., Harayama, K., and Bamba, T. (2018). Development of an analytical method for polycyclic aromatic hydrocarbons in coffee beverages and dark beer using novel high-sensitivity technique of supercritical fluid chromatography/mass spectrometry. *J. Biosci. Bioeng.*, 126, 126–130.

[68] Evans, A. M., DeHaven, C. D., Barrett, T., Mitchell, M., and Milgram, E. (2009). Integrated, nontargeted ultrahigh performance liquid chromatography/electrospray ionization tandem mass spectrometry platform for the identification and relative quantification of the small-molecule complement of biological systems. *Analytical chemistry*, 81(16) 6656-6667.

[69] Angeline, N., Lemasson, E., Mauge, F., Petit, A., Bertin, S., Hennig, P., and West, C. (2019). Purification of drug degradation products supported by analytical and preparative supercritical fluid chromatography, *J. Pharm. Biomed. Anal.*, 40–47. https://doi.org/10.1016/j.jpba.2019.03.033.

[70] Schiavone, N. M., Bennett, R., Hicks, M. B., Pirrone, G. F., Regalado, E. L., Mangion, I., and Makarov, A. A. (2019). Evaluation of global conformational

changes in peptides and proteins following purification by supercritical fluid chromatography, *J. Chromatogr. B Anal. Technol. Biomed. Life Sci.*, 94–100. https://doi.org/10.1016/j.jchromb.2019.02.012.

[71] Karila, L., Megarbane, B., Cottencin, O., and Lejoyeux, M. (2015). Synthetic Cathinones: A New Public Health Problem, Curr. *Neuropharmacol.*, https://doi.org/10.2174/1570159x13666141210224137.

[72] Hohmann, N., Mikus, G., and Czock, D. (2014). Effects and Risks Associated with Novel Psychoactive Substances: Mislabeling and Sale as Bath Salts, Spice, and Research Chemicals, Dtsch. *Arztebl. Int*, https://doi.org/10.3238/arztebl.2014.0139.

[73] Prosser, J. M., and Nelson, L. S. (2012). The Toxicology of Bath Salts: A Review of Synthetic Cathinones, *J. Med. Toxicol*, 33–42. https://doi.org/10.1007/s13181-011-0193-z.

[74] Carnes, S., O'Brien, S., Szewczak, A., Tremeau-Cayel, L., Rowe, W. F., McCord, B., and Lurie, I. S. (2017). Comparison of ultra high performance supercritical fluid chromatography, ultra high performance liquid chromatography, and gas chromatography for the separation of synthetic cathinones, *J. Sep. Sci.*, 3545–3556. https://doi.org/10.1002/jssc.201700349.

[75] Lübeck, J. S., Malmquist, L. M. V., and Christensen, J. H. (2019). Supercritical fluid chromatography for the analysis of oxygenated polycyclic aromatic compounds in unconventional oils, *J. Chromatogr. A.*, 1589, 162–172. https://doi.org/10.1016/j.chroma.2018.12.056.

[76] Orecchio, S., and Papuzza, V. (2009). Levels, fingerprint and daily intake of polycyclic aromatic hydrocarbons (PAHs) in bread baked using wood as fuel, *J. Hazard. Mater.*, 164, 876–883. https://doi.org/10.1016/j.jhazmat.2008.08.083.

[77] Yoshioka, T., Nagatomi, Y., Harayama, K., and Bamba, T. (2018). Development of an analytical method for polycyclic aromatic hydrocarbons in coffee beverages and dark beer using novel high-sensitivity technique of supercritical fl uid chromatography/mass spectrometry, *J. Biosci. Bioeng.*, 126, 126–130. https://doi.org/10.1016/j.jbiosc.2018.01.014.

[78] Van zelst, F., van meerten, S., and Kentgens, A. (2019). *Characterising polar compounds using supercritical fluid chromatography - nuclear magnetic resonance spectroscopy (SFC-NMR)*, 219–232. https://doi.org/10.1039/C8FD00237A.

[79] de Kock, N., Acharya, S. R., Ubhayasekera, S. J. K. A., and Bergquist, J. (2018). A Novel Targeted Analysis of Peripheral Steroids by Ultra-Performance Supercritical Fluid Chromatography Hyphenated to Tandem Mass Spectrometry, *Sci. Rep*, 1–9. https://doi.org/10.1038/s41598-018-35007-0.

Chapter 5

Capillary Electrophoresis: Advancements and Applications

**Sumit Pasricha*
and Sanskar Ahlawat**
Department of Pharmaceutical Analysis, ISF College of Pharmacy, Ghal Kalan, Punjab, India

Abstract

A method for isolating ions based on their electrophoretic mobility in the presence of an electric field is capillary electrophoresis. Ion charge, viscosity, and radius all have an impact on electrophoretic mobility. In this chapter, we have discussed various types, methodologies, factors affecting, applications and some advancements of capillary electrophoresis. The use of capillary electrophoresis in different segments of industry like biopharmaceuticals, microfluidic chip electrophoresis, forensic analytical chemistry, agrochemicals, separation of vitamins and minerals, chiral analysis, bio molecular interactions and cancer studies. In hyphenation with CE-MS and CE- NMR, which produce high efficacy, high throughput and less time-consuming technique. CE-MS is a hyphenated technique which separate combined form of liquid by capillary electrophoresis and mass spectrometry. In CE-NMR, the separation occurs with nuclear magnetic resonance.

Keywords: capillary electrophoresis, mass spectrometry, nuclear magnetic resonance, electrophoretic mobility, chromatographic technique

---

* Corresponding Author's Email: pasricha196@gmail.com.

In: Modern Approaches in Fluid Chromatography
Editors: Rohit Bhatia and Bhupinder Kumar
ISBN: 979-8-88697-856-8

1. Introduction

Capillary electrophoresis is a technique for isolating ions based on their electrophoretic mobility in the presence of an electric field. Radius, viscosity, and ion charge all affect electrophoretic mobility [1]. The velocity of the particle depends on the voltage being used. In these steady-state conditions, the electric field strength is inversely linked to the ion's velocity. The electrophoretic mobility serves as the proportionality constant [2]. The mobility of distinct molecules varies depending on their average charge, size, and shape, as well as the qualities of the solvent medium [3]. This phenomenon serves as the foundation for electrophoretic separation procedures reaches to constant. Capillary electrophoresis is the heart of electroosmotic flow and electrophoretic mobility. It is an analytical method used to separate their ions according to their electro-osmotic flow and electrophoretic mobility using high voltage. It is a technique that separates very small quantities of the substance in a short time with very high voltage. It describes the movement of ions or molecules under the influence of applied voltage [4].

In the year 1823, the Lodge invented the migration of phenolphthalein jelly in a tube and in 1937 Tiselius invented moving boundary electrophoresis. In the year 1979, Mikkers et al. found polymer capillary electrophoresis [5]. In 1989, Arnold Orville Beckman was first introduced as an analytical instrument of capillary electrophoresis. He also invented the modern-day PH meter. In 1991, they are introduce the hyphenated technique LC-CE [6].

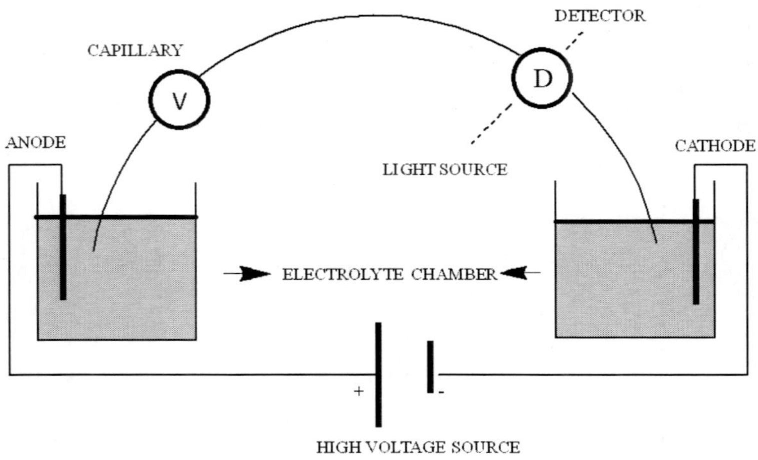

Figure 1. Layout for instrumentation of capillary electrophoresis.

1.1. Principle

When an external electric field is given to a solution of charged ions, the ions move in the opposite direction of the electrodes, which is the basis of the electrokinetic separation process. Many molecules in a solution are charged ions, either positive or negative ions. The frictional drag barely stabilizes the electric force as the charge of motion quickly reaches a persistent amount (Figure 1) [1].

2. Working of Capillary Electrophoresis

2.1. Migration Rate of Ions in CE

The electric field's intensity has a direct correlation with the rate at which ions migrate (v). It is completely dependent on an electric field that is inversely proportional to its length L and directly proportional to the applied voltage V [7].

$$v = ue * V/l \qquad (1)$$

It indicates higher voltage is applied desired to archive rapid ionic separation take place. It is advantageous to have rapid separation, but still more important is to attain high-resolution separation [8]. So, we must scrutinize the component that determines resolution in electrophoresis [9].

2.2. Plate Height in Capillary Electrophoresis

In this chromatography, both mass transfer resistance and longitudinal diffusion resistance are put up to band broadening [10]. However, because only a single stage is used in electrophoresis, in theory only longitudinal diffusion is necessary. But in the practical aspect, a joule heating effect can add variance as well as its injection process. Although capillary electrophoresis is not a chromatographic technique it is similar to a chromatographic technique [11]. So, we can calculate plate count by this equation:-

$$N = ueV/2D \quad (2)$$

D = solute diffusion coefficient (cm^2s^{-1}). Because resolution increase is directly proportional to plate count increase, it is advantageous to use high applied voltage to attain high-resolution separation [12].

2.3. Electroosmotic Flow

Electroosmotic flow is begun by the electrical double layer that progresses at the silica resolution's interface, (Figure 2) [13]. At that point, the pH value is more than 3, and a silica capillary's interior wall has a negative charge to the surface silicon group being an ionized negative surface of the silica capillary is close to an electrical double layer of buffer cations [14].

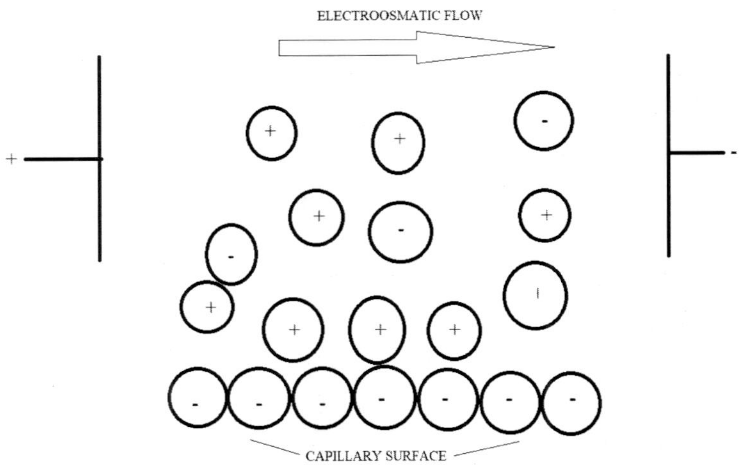

Figure 2. Electroosmotic flow of capillary electrophoresis.

The diffuse outer layer of the solvated double-layer cations is drawn to the cathode or the negative electrode and pulls solvent along with it.

The diffuse outer layer of the solvated double-layer cations is drawn to the cathode or the negative electrode and pulls solvent along with it [15]. The electroosmotic flow essentially serves as the mobile-phase pump of capillary electrophoreses. Since it often moves at a rate faster than the individual ions' electrophoretic migration velocities [16]. Even when analytes move across the capillary according to their charges, all positive, neutral, and even damaging species can typically be swept toward the same end of the capillary by the electroosmotic flow rate (Figure 3) [17], allowing for their detection as they pass by a common point. The final electropherogram resembles a chromatogram but has more defined peaks [18].

Figure 3. Comparison of electroosmotic flow and laminar flow.

2.4. Buffer Species

Buffers are chemicals that control the pH of a solution. They are often weak bases or acids that can accept or even provide protons. Buffers reduce the pH shift that occurs when more acid or base is added, and CE buffers are selected depending on the required pH range as well as other factors [19].

Different types of buffers have been employed in CE systems. The most often utilized ones include citrate, acetate, phosphate, tris (trishydroxy-methylamino methane) and borate [20]. The zwitter ionic buffers have an advantage over mono-functional buffer molecules in that they carry less current [21]. When choosing a buffer for CE, there are a number of considerations into account, as well as the optimum operating temperature, charge pH value of the buffer with respect to the analytes and capillary wall, and impacts on detection [20].

2.5. Buffer Concentration

The capacity of the system to buffer an addition of acid or base will improve with an increase in buffer concentration, reducing the possibility that the system's pH may alter. The concentration of the buffer can also have an impact on other elements of CE. When attempting to attain the best resolution and reproducibility, these elements may be crucial [22].

The stacking phenomenon may happen if the sample plug's composition differs from the buffer's composition inside the capillary [23]. Sharper peaks and greater sensitivity may arise from properly utilised stacking. To achieve stacking, keep the conductivity of the sample below that of the running buffer by lowering the concentration of the sample buffer in relation to the run buffer. To understand this phenomenon, picture a wire as a buffer-filled capillary [24]. If the buffer concentration is constant from one end of the capillary to the other, the voltage drop will be linear [25]. A resistor would be inserted into a wire if a plug of the diluted sample were to be inserted. Now, the area around the plug experiences a steeper voltage drop than the surrounding areas [20]. The analyte's velocity will be higher in this area since it is proportional to the voltage gradient. Analytes will move swiftly up to the border between the less concentrated running buffer and the diluted sample plug. Once there, they will stack and create a zone that is smaller than the initial injection plug, increasing the analyte concentration there [26].

3. Instrumentation of Capillary Electrophoresis

Capillary electrophoresis uses relatively straightforward equipment. A fused silica capillary filled with buffer typically measures 10-100 um in internal diameter and is 30-100 cm long [27]. Two buffer reservoirs are also used to store platinum electrodes. The outside walls of fused-silica capillaries are commonly covered with polyimide for strength, flexibility, and stability. A sample is inserted at one end, while detection occurs at the other [28]. Any capillary electrophoresis system must have a high-voltage power source, a detector, a capillary, and a safety unit to shield the user from the high voltages used in the process [29]. The adoption of this method in pharmaceutical analysis has been promoted by the availability of commercial CE systems, which have improved injection and migration repeatability as well as detection [30].

3.1. Sample Inducer

The injection of large amounts is impossible because of the small dimensions associated with CE [31]. Electrokinetic and displacement injection are the two most popular sample introduction techniques. For sample injection, the majority of commercial devices use a pressure differential [32].

With an electrokinetic injection, the capillary electrode and one end was taken out of the buffer compartment and after placed the sample in a cup after that, a voltage is given for a pre-determined period, allowing the sample to move electro-osmotically and migrate ionically into the capillary [33]. When the separation is complete, the capillary end and electrode are re-inserted into the standard buffer solution [34].

In displacement injections, an imperceptible cup containing the sample is likewise used as the sample introduction end of the capillary, but this time, a pressure differential forces the sample solution into the capillary [35]. By providing a vacuum at the detector end, pressurizing the sample, or raising the sample end, the pressure difference can be created. Because of ion mobility, pressure injection does not discriminate, however, it cannot be employed in capillaries filled with gel [36].

The time of the injection determines the volume injected for both electrokinetic injection and pressure injection. Injections of 5 to 50 nL are typical, while volumes below 100 pL have been reported [37]. A height differential of 5 cm for a 10 second injection of roughly 6 nL with a 75 μm inside diameter [38].

3.2. Capillary Technology

The fused silica capillaries used in capillary electrophoresis are flexible because of the exterior polyamide covering [39]. Additionally, they come with coatings varying from 10 to 300 μm in inner diameter. Because fused silica is UV light transparent, the capillary can function as a standalone detection flow cell. The large surface-to-volume ratio of these capillaries, which enables the active dissipation of heat generated by high voltages while maintaining efficiency separations [11], is primarily responsible for the success of capillary electrophoresis. Due to the high probability of solute surface interactions, this high surface-to-volume ratio in capillary electrophoresis also presents a problem. In the case of basic proteins/peptides, coulombic interactions with the capillary surface are particularly problematic and can affect reproducibility, efficiency, and in severe circumstances, elution itself [40]. Several of our employees advise the usage of the surface coating to lessen or even completely eliminate electro-osmotic flow in capillary electrophoresis in addition to reducing the adsorption of analytes [41]. The hydrolytic stability over a wide pH range and the repeatability of chemically altered surfaces are issues that concern those who exploit these capillaries. Even though polymer coatings typically exceed their nonpolymeric counterparts (Figure 4), the problem of consistently manufacturing polymerized surface coatings from batch to batch is addressed. Silanization reactions are challenging to replicate for a variety of reasons, according to Schomburg [42].

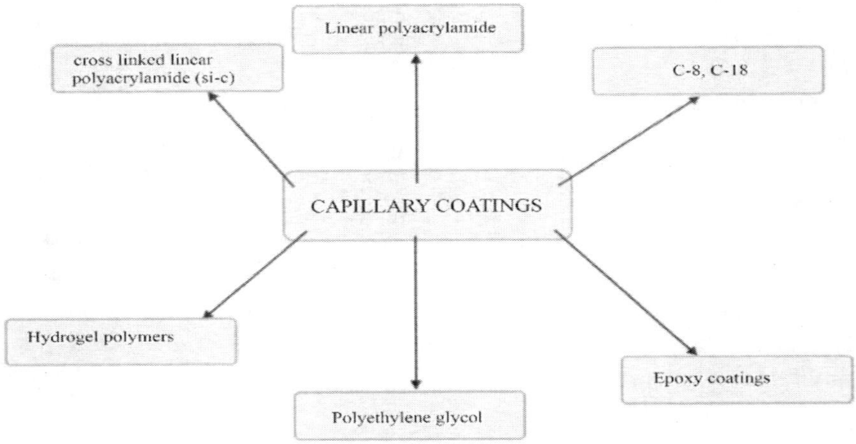

Figure 4. Types of different capillary coatings.

3.3. Detection

Various types of significant development occurred in detector design and capillary electrophoresis [38]. The development of a more effective therapeutic creature as well as increased efficiency and the requirement to detect the small amount of the active drug and associated contaminants. Any capillary electrophoresis system must have a detector, a capillary, a high-voltage power supply, and a safety unit in order to safeguard the user from the high voltages used throughout the operation [43]. The electrophoretic counter electrode serves as a reference for the detector as well. A 90 mm platinum wire that acted as the electrode body was passed through a 100 mm fused silica capillary to create the cunning electrodes [44].

3.3.1. UV Absorption Detection

The most prevalent method for capillary electro-phoresis used in the detection and other connected chromatographic techniques is UV absorption [45]. The key aspects of its popularity are its universality and broad HPLC application.

The ability to identify components up to 95% is what makes the quantity of light absorption in the UV or visible area by far the best detection technique in CE.

The detection can be directly done by capillary, without any direct contact between any separation channel and measurement system [46]. In fact, in this method, the capillary itself can act as a perfect optical cell because of its building material. The capillaries in the CE are made up of fused silica (quartz-like material) which let wavelengths down to 190 nm pass through itself [47].

Almost 65% of samples are analyzed by liquid chromatography at 254 nm wavelength and 90% of samples are analyzed by a variable wavelength detector. There has a broad range of samples that should be detected by UV absorbance with capillary electrophoresis [48].

In the UV absorbance system, the magnitude of capillary electrophoresis is much higher than the HPLC [49]. But the major disadvantage is the detection of a component on a different wavelength. In fact, in capillary electrophoresis, the detection of different samples takes place at less wavelength [50].

3.3.2. Fluorescence Detection

The second largest capillary electrophoresis detection method is based on fluorescence, which can be excited by a lamp or a laser [51]. It is exclusively utilized in biological applications for analytes that comprise all primary amines, including amino acids, peptides, and proteins. This highly sensitive and selective detector is essential for DNA sequencing. Compared to UV absorbance, the magnitude of the limit of detection is smaller for fluorescence. However, due to the relatively small dimensions of capillary electrophoresis, it is believed that detection sensitivity may be questionable for many applications [52]. In particular, for trace analysis, it appears doable to improve or alter a fluorescence detector for use in CE (Figure 5) [53].

A fluorescence detector created by Jorgenson and Lukas concentrates the required excitation wavelength onto a particular region of the column using a high-pressure mercury arc lamp and glass filters [54]. It updated the detector's design by renovating the filter system for excitation wavelength confinement with a double monochromator. This improved wavelength discrimination, decreased stray light levels, and increased excitation wavelength adaptability. The diagram below shows how the column fluorescence detector's optical setup is set up. The range between 10-7-10-6 M was the system's detection threshold [55].

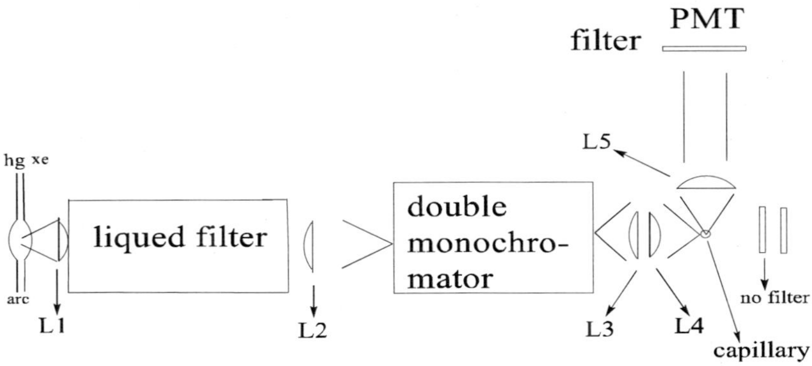

Figure 5. Schematic diagram of the fluorescence detector.

3.3.3. Electrochemical Detection

Amperometric and conductivity detection, which are based on the dimension of an electric current, will be discussed in this chapter. Both anticipate that the separating medium and the sensing element(s), the electrode(s), will come into a direct brush [56].

A high-pressure mercury arc lamp and glass filters are used in a fluorescence detector created by Jorgenson and Lukas to concentrate the intended excitation wavelength onto a particular region of the column [57]. It improved wavelength discrimination, decreased stray light levels, and increased excitation wavelength adaptability by reinstating the filter system for excitation wavelength confinement with a double monochromator. The diagram below shows how the column fluorescence detector's optical setup is set up. The detection threshold for the system was in the range of 10-7-10-6 M [58].

In CE-ED, there are three possible locations for the detecting electrode(s). In a process known as online mode, the sensing electrodes can be positioned anywhere along the capillary's height [59]. This is clearly the usual mode for spectrometric detection, despite the fact that there is no direct connection between the sensor and the solution in that position. For so called end-column detection, a second possibility is to segment the electrode(s) straight into the buffer vial at the capillary's end, together with the grounding electrode of the separation systems. Off-column detection is the third option. In this method, detection occurs after the electrophoretic voltage has been decoupled [60]. Following decoupling, the resolution containing the distinct analyte zones is transferred in the direction of the detector and terminated by the second capillary pieces (Figure 6) [61].

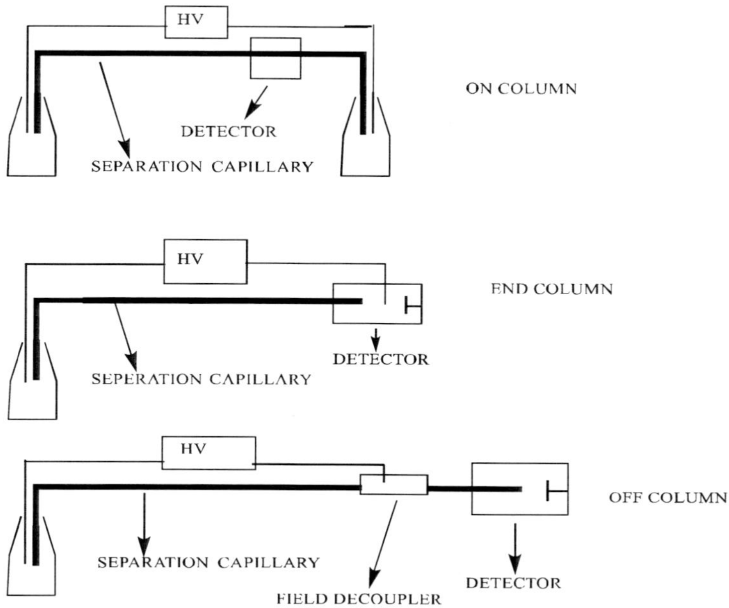

Figure 6. Working of electrochemical detector.

3.3.4. Mass Spectrometry Detector

With the use of mass spectroscopy, a substance of structural data may be gleaned from a small amount of material. Capillary electrophoresis and mass spectrometry can be combined utilizing continuous flow, rapid atom bombardment techniques, or electron spray ionization. When electropherograms generated by the contaminated sample hide solute identification, several peaks become a serious problem [62]. Credential difficulties can be resolved by combining CE with mass spectrometry (MS). The relatively few analytes that elute from an electrophoresis capillary may best be studied with this technique. There is a benefit of connecting CE with MS, comparable to HPLC. This benefit arises from the fact that the MS's capacity to remove eluent at moderate surge rates and with tiny volumes of supplies eluting after the capillary is compatible [63]. The two organizations and the removal of the electrolyte system have been the main barriers to the expansion of CE MS [62].

3.3.5. Raman Based Detector

A distinct spectrum of an analyte with a very high spectral resolution is provided by Raman spectroscopy (RS). It has proven to be a highly helpful technology thus far, however, it is rarely employed for detection in CE [64]. This is due to the fact that, as compared to the fluorescence/absorbance method, Raman emission is a very inefficient process that results in poor sensitivity in detection [65]. Raman spectroscopy is an appropriate detection technique for CE since it can be used to collect qualitative information about analytes from spectra. With this technique, the analyte is exposed to a monochromatic light source and the frequency and intensity of the dispersed light are measured [66]. The appeal of Raman spectroscopy, on the other hand, is multifaceted [67].

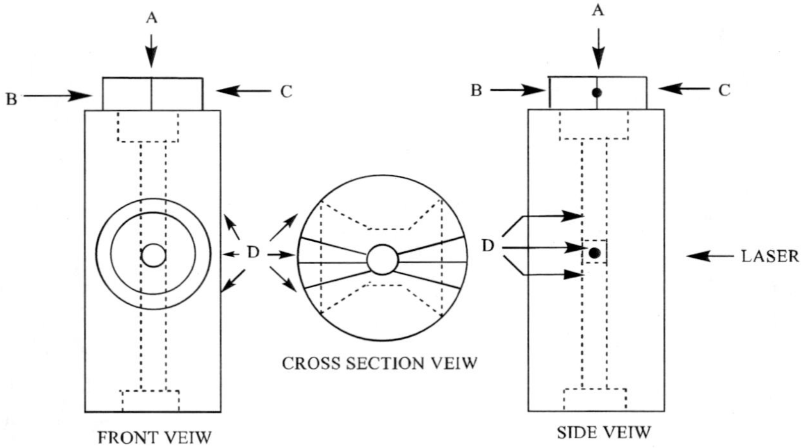

Figure 7. A: threaded cap, B: set screw, C: capillary bore, and D**:** holes for optical fibers.

In the first place, Raman spectra have long been an amusing source of information about chemical bonding in the system of vibrational bands exclusive to different molecular structures. Second, it can be recorded at any excitation frequency, with different fluorescence. Therefore, a perfect wavelength excited can be selected to minimize undesired photochemical effects or related fluorescence [68]. Moreover, its detection may be simply integrated with capillary electrophoresis by making a few changes to the present system used for optical detection approaches. Chen and Morris's demonstration of online Raman detection for CE offers a lot of potential for locating and measuring tiny compounds (Figure 7) [69].

4. Method and Modes of Capillary Electrophoresis

4.1. Capillary Zone Electrophoresis

Liquid phase separation is a method used in zone electrophoresis in capillaries. It consists of several different sub-techniques, such as capillary gel electrophoresis, micellar electrokinetic capillary chromatography and capillary iso-electro focusing. It has been separated by both small and large molecules [70]. A strong potential field is carried by capillary zone electrophoresis, which produces a highly effective ion separation (Figure 8) [71].

Capillary zone electrophoresis is used to detection of new developments. The detection scheme directly or indirectly depends upon UV absorption, fluorescence, radioisotope, mass spectrometer, and electrochemical detectors [72]. The mostly electrochemical detection scheme is used in capillary zone electrophoresis because it involves conductivity measurement and amperometry [73]. It also elaborates on the column and post-column put a stop to high potential separation used the form to impede the electrochemical process. One strategy involves the establishment of a 40 μm diameter hole in a capillary with the help of a laser [74]. Small platinum wires are then inserted into the perforations to verify the detection of column conductivity. Additionally, it describes how the column and post-column employed the shape to obstruct the electrochemical process and prohibit high potential separation [75]. An additional technique is to a capillary fracture with porous glass to permit off-column amperometric detection. The challenge lies in how to generate such methodically and cost-effectively [76]. This methodology works for both direct and indirect electrochemical detection. This restriction applies to both electrochemical detection methods for capillary zone electrophoresis [77].

Another sort of capillary zone electrophoresis takes place when a microelectrode for conductimetric and amperometeric detection is positioned at the capillary's output. These end-column detectors are simple to put together [78]. They are not negatively affected by the high electrode potential that capillary zone electrophoresis separation produces [79].

4.2. Capillary Gel Electrophoresis

The interpretation of separation efficiency from classical gel electrophoresis was updated and experimentally accepted for capillary gel electrophoresis [80]. The distinct gene products can be recognised and identified. A method for studying the feature of capillary gel electrophoresis that is comparatively simple, rapid, and exceedingly sensitive [81]. The theory of electrophoresis states that when an electric field is supplied, usually using stuck electrodes, charged molecules will move across a matrix [82].

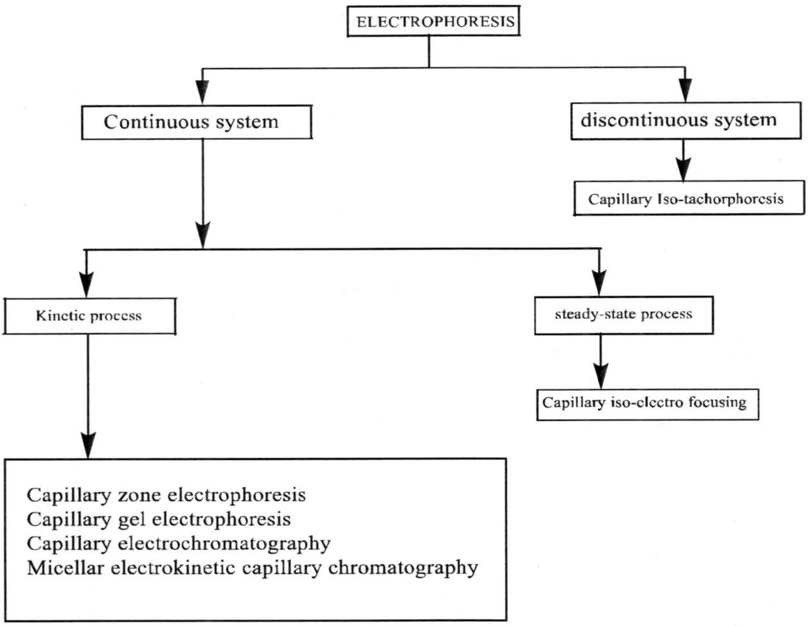

Figure 8. Modes of capillary electrophoresis.

Capillary gel electrophoresis (CGE), which allows for high-resolution separation, is a rapidly developing technique [83]. Slab gel and capillary gel electrophoresis are fundamentally distinct from one another. Slab gel electrophoresis permits even analytes with low sufficient mobilities to be separated and visualised, in contrast to capillary gel electrophoresis, which requires all solutes to drift after the injection point to the detection cell in order to be observed [84].

Gels are presumably advantageous for electrophoretic separations because they enable separation based on "molecular sieving" [84]. They also function as anti-convective media, lessen solute diffusion, which widens the zone, inhibit solute adsorption to capillary walls, and provide consolation in preventing electro-osmosis [85]. To be an effective electrophoretic medium, the gels must maintain a few certain characteristics, including temperature stability and the appropriate range of pore size. Since this procedure constrained electro-osmotic flow, the technology is limited in that neutral molecules cannot pass through the gel (Figure 8) [86].

4.3. Capillary Electrochromatography

In recent years, capillary electrochromatography—a mix of CE and HPLC—has gained prominence. Capillary electrochromatography, the solutes are moved through a capillary filled with stationary phase material using an electrically difficult flow. Separation may result from differentiation [87].

In the chromatographic separation method known as capillary electrochromatography, an electrical field is employed to induce electro-osmotic flow (EOF) between a stationary phase and the mobile phase [88]. In the separation column, a fused silica (FS) capillary tube, 50–200 lam I.D., is frequently active. It is filled with stationary phase particles of the reversed-phase HPLC kind. The particles are kept in the packed bed by taking precautions [89]. To impact separation, solvents like in HPLC are utilised, such as acetonitrile-aqueous buffer combinations. For the EOF to move at a speed that is appropriate, there is an applied electrical field of 1000 V/cm [89].

The resolution domain of CEC might be equivalent to capillary gas chromatography (GC), which is currently not improved by HPLC, with smaller particle size, longer columns, and cork flow. In contrast to HPLC, where selectivity is frequently the basis of separation, CEC makes it possible to separate complicated assortments by making use of the higher plate numbers that are readily available (Figure 8) [90].

4.4. Micellar Electrokinetic Chromatography

Solutions using a basic buffer that is non-aligned and cannot be intended by CE. Thus, in MEKC, neutral components are chromatographically separated using micellar surfactant solutions [91]. When the surfactant concentrations are high enough, the molecules of the surfactant carefully group to form micelles. In a chromatographic pattern, non-aligned solutes partition with these micelles, similar to reversed-phase HPLC, which are separated by retention factor [92]. At satisfactorily high surfactant absorptions, Non-aligned solutes partition with these micelles in a chromatographic pattern. Similar to reversed-phase HPLC, neutral solutes separate from these micelles according to their retention factors after partitioning with them in a chromatographic trend [93].

MEKC can be carried out using a standard CE instrument like CZE with the simple use of a micellar solution. To make the micellar solution, an ionic surfactant is dissolved in a buffer at a concentration higher than the critical micelle concentration (CMC) [94]. The critical micelle concentration, at which the surfactant concentration and the micelle are in equilibrium (CMC), is equivalent to the surfactant concentration at which the surfactant forms the micelle [95]. Several methods, including electrical conductivity, cyclic voltammetry, surface tension, spectrofluorimetry light scattering, spectrophotometry, nuclear magnetic resonance, and CE, can be used to measure it.

Since different micelles exhibit variable selectivity, one benefit of MEKC is the simple accessibility of multiple surfactants. Bile salt micelles, for instance, provide dramatically different selectivity. Because of their chirality, bile salts are also helpful for enantiomeric separation (Figure 8) [96].

4.5. Capillary Iso-tachorphoresis

To separate these two buffers a sample plug was added via this technique. During the separation, the predominant electrolyte among them moves the most. The mobility of the second electrolyte, the trailing electrolyte, is the lowest of all [97]. Both the analytes and the buffer's charge indication must match. The ions in the sample form discrete zones that are all near to one another and do not produce peaks when the voltage is applied [98]. Within a zone, the analyte concentration is constant, and the zone's length is proportional to its concentration. Because of the homogeneous voltage drop inside a zone, iso-tachophoretic methods and conductance detectors complement one another nicely. Protein and peptide capillary iso-tachophoresis has been employed with MS detection [99]. This method of separation may be the optimum separation mechanism for this detection method since, it makes continual concentration zones visible to the MS (Figure 8) [100].

4.6. Capillary Iso-electro Focusing

Molecules having both positively and negatively charged groups exhibit an equal number of positive and negative charges at a certain pH level. At particular pH, the molecule behaves as though it is neutral despite being charged since the positive and negative charges balance each other out [40]. This property is known as the iso-electric pH or pi. The molecule has no tendency to move in an electrical field as a result. A pH gradient is created inside the capillary by the speciality chemicals known as ampholytes [101], which are utilized in iso-electric focusing. These ampholytes are mixtures of several buffers with pKa values. Ampholytes will assemble themselves according to their pKa values in an electrical field; this gradient is stuck between two strong acids and bases. When an analyte is added to this gradient, it will migrate to the region where its pH is equal to its pi [102]. The analyte stops migrating at this moment because it has no net charge. As long as the pH gradient is stable and the voltage is provided, it will stay in that position [103].

It is frequently used to investigate the distribution of glycoprotein and carbohydrate isoforms utilizing capillary iso-electric focusing. These proteins typically behave differently in other methods, such as Sodium dodecyl gel electrophoresis or Capillary gel electrophoresis [104], where they move as diffuse bands and produce large peaks. These bands can frequently be resolved into peaks by capillary iso-electro focusing with differences as small as one charged sugar group (Figure 8) [105].

5. Factor- Effecting Capillary Electrophoresis

When doing electrophoretic separation, zone dispersion and Joule heating are important considerations, regardless of whether it is carried out in a standard slab, tube, or microcapillary format [106]. However, the intense electric fields, the high surface-to-volume ratio, and the low viscosity of the counter electrode become particularly significant in capillary electrophoresis (CE) separations. The eventual separation performance will depend on how well these effects are managed. The ideas described here will be referenced throughout the chapters of this volume because they are essential to the execution of all CE separations.

5.1. Joule Heating Effects

The capacity to disperse the Joule heat created by the electric current transient across the electrophoresis buffer severely limits the speed, precision, and scale of electrophoretic separations. A variety of factors, including the temperature differential that emerges from this Joule heating, might compromise the separation quality [107].

First, it is initially possible to create density gradients in the electrophoresis buffer when the temperature gradients are extreme enough. Which can subsequently result in spontaneous convection. Convection of this kind would mix up previously separated sample zones, significantly degrading the effectiveness of separation [108]. Second, introducing a spatial dependency to electrophoretic mobility. This is a strong temperature-dependent variable, that may have an impact on separation performance even if the temperature gradients aren't large enough to trigger natural convection. This spatial dependency on mobility may lead to a distortion in the migration zones, which would impair separation performance [109]. Third, if the average temperature of the buffer noticeably rises, the organisational integrity of the sample analyte may be conceded.

5.1.1. Natural Convection

The bulk flow of fluid resulting from variations in densities between adjacent fluid phase areas is known as natural convection, also referred to as free convection. Changes in composition or temperature may be the root cause of these density variations. Here, variations in density brought on by temperature variations were of interest.

The dimensionless Rayleigh number, Ra, is one indicator that the impacts of natural convection can be correlated with (Cussler, 1984).

$$Ra = R_1{}^4 g/\eta\alpha\ (\Delta p/\Delta r) \quad (3)$$

Here, R= radius for electrophoresis chamber, which in this case is a tube, g = gravity of acceleration, the medium viscosity is 77, It is the medium thermal diffusivity, and Ap/Ar is the change in density caused by heating per unit of radial distance. Based on theoretical projections, natural convection will become significant when Ra is on the order of 1, and it will be incredibly feeble when Ra is less than 1. As a result, The Rayleigh number is a numerical standard that can be used to assess the likelihood of spontaneous convection [110].

5.2. Intracapillary Temperature Profile

The energy-balance equation must be resolved in order to estimate the temperature gradients brought on by Joule heating inside the capillary. According to this equation, the difference between the thermal energy input and departure rates equals the amount of energy build-up in any control volume. Use the relation to represent this in cylindrical coordinates.

$$Se = 1/r\ d/dr\ (rq_r) \qquad (4)$$

Where qr is the radial position of r and Se is the flow of the generator with a control per unit volume. It is the rate of power production per unit volume inside the control volume. Se is provided by in the case of Joule heating.

$$Se = I^2/k_e \qquad (5)$$

Here, I is the current density and electrical conductivity of the electrophoresis buffer is Ke. It simplifies the study to providing a more convenient equation, we assume in this derivation that Ke is constant across the whole conducting medium. This just means that we may ignore Ke variation inside the capillary, not that Ke's absolute value is stable. For the majority of CE situations, it has been demonstrated that this assumption results in minimal error [111].

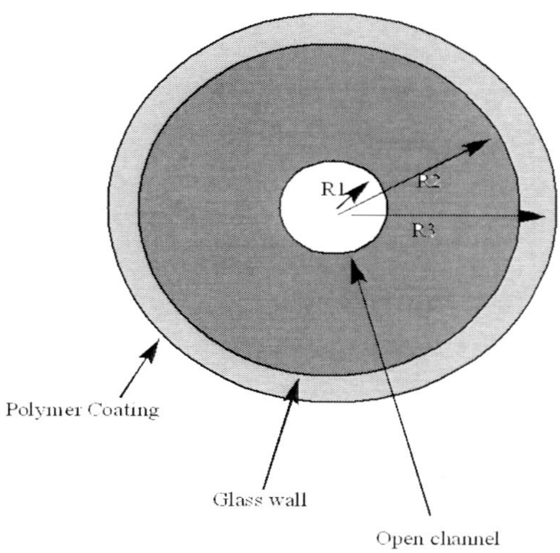

Figure 9. Diagram of a capillary tube, Where the radius of the glass wall, open channel and polymer coating, in that order, are R1, R2, and R3 (Figure 9).

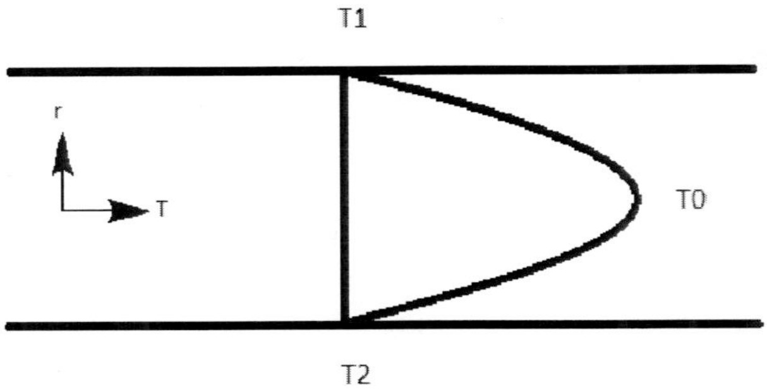

Figure 10. Inside the capillary, the temperature profile is parabolic. T0 is the temperature at the capillary center-line, and 7 is the temperature at R1.

Result of the equation is r = 0

$$q_r = Se\ r/2 \quad (6)$$

To define the previous equation in terms of T instead of qr, we utilise the Fourier law of heat conduction [112], which asserts that since with r, the heat flux rises linearly.

$$q_r = -k_b\ dt/dr \quad (7)$$

Where the thermal conductivity of buffer is k_b, from equation 6,

$$-k_b\ dt/dr = Se\ r/2 \quad (8)$$

Assuming that k_b is constant throughout the capillary, using the boundary condition, it is simple to integrate the aforementioned equation that at $r = R_1$, $T = T_1$. This results in the expression.

$$T(r) - T_1 = SeR_1^2/4k_b\,(1-(r/R_1)^2) \quad (9)$$

In (Figure 10), a parabolic temperature profile can be calculated with the equation above. This profile will be used later when we talk about how it affects zone dispersion [113]. Simply evaluate the equation above at r = 0 to obtain the overall temperature differential throughout the buffer, T_0-T_r.

$$T_0 - T_1 = SeR_1^2/4k_b \quad (10)$$

Where, ro is the capillary centre-line temperature. This expression does not tell about the overall size of the capillary temperature increase relative to the surrounding temperature, Ta. It only discusses the variation in temperature between the interior wall and the capillary centre [114].

5.3. Overall Temperature Rise

It is essential to regulate the total temperature increase within the electrophoresis buffer below a certain recognized operating condition. To get the total temperature increase, TQ - 7a, where Ta is the thermal gradient of the adjacent medium, it has to be resolved inside the tube wall, within the polymer coating, and within the ambient air. Performing the same thing as previously within the glass wall.

$$T_1 - T_2 = SeR_1^2/2k_g \ln (R_2 - R_1) \quad (11)$$

And within the polymer coating,

$$T_2 - T_3 = SeR_1^2/2k_p \ln (R_3/R_2) \quad (12)$$

Where T_2 and T_3 are the temperatures at locations R_2 and R_3, respectively, and kp stand for the glass's and the covering's relative thermal conductivities. Se_{tot} order to T_3 to T_a, they adhere to the constraint that the rate of heat generation inside the capillary must match the total energy leaving it at the capillary-air contact [115]. Consequently, given that/z represents the capillary-air heat transfer coefficient,

$$ha(T_3 - T_a) = Se_{tot} \quad (13)$$

Where A represents the capillary's outside surface area. Consequently, substituting for A and the capillary's volume,

$$T_3 - T_a = SeR_1^2/2R_3h \quad (14)$$

Therefore, to calculate $T_0 - T_a$, we simply add equations (10,12,14)

$$T_0 - T_a = SeR_1^2/2\ (1/2k_b + 1/k_g \ln (R_2/R_1) + 1/k_p \ln (R_3/R_2) + 1/R_3h) \quad (15)$$

According to equation 4, this is the contact we have been pursuing. It links the rate of power generation and, subsequently, the strength of the electric field to the rise in temperature inside the capillary, in addition to the features of the capillary, the buffer, and the surrounding medium. It's essential to understand that, in the large majority of cases of practical interest, the last term dominates the overall temperature rise, $T_0 - T_a$. In equation 4,

$$T_0 - T_a = SeR_1^2/2R_3h \quad (16)$$

This is due to the fact that the impedance to heat transfer between the capillary's external wall and its surroundings is usually far higher than the total of all other resistances [116].

6. Applications of Capillary Electrophoresis

6.1. Chiral Capillary Electrophoresis

In chiral capillary electrophoresis, separation methods are extremely effective instruments for chiral analysis [117]. Although LC has been the most extensively employed method, CE has demonstrated to offer some extremely consider going for doing enantiomeric separations. On the one hand, several of the inherent qualities of CE, such as its high efficiency and resolution, low consumption of reagents and samples [118], and quick analysis periods, are advantageous to chiral analysis [119].

On the other hand, the ability to add the chiral selector directly to the separation medium in the so-called Electrokinetic Chromatography (EKC) mode gives chiral CE a high degree of flexibility because the type of selector (or combination of chiral selectors) and its concentration can be changed with ease [120], facilitating the separation of enantiomers and lowering the economic costs associated with the use of chiral chromatographic columns [121].

6.2. Bio Molecular Interactions

Capillary electrophoresis (CE) is fetching a more well-known analytical technique in medicine and biology [41]. With its quick processing times, great resolution, and minimal sample and reagent use, CE is a popular technology for drug screening and mass bioassays. It is a fast reversible biomolecular interaction in the developing field. That is characterised as extremely selective biomolecule-ligand non-covalent binding [122]. These interactions on affinity regulate immunological response, DNA replication, gene expression, cell recognition, signal transmission, and other cellular processes. To appreciate biological processes for these reactions under capillary electrophoresis govern, it is required to know the quantitative properties of binding reactions (equilibrium or rate constants) [123].

6.3. Capillary Electrophoresis in Biopharmaceutics

Proteins, liposaccharides, glycoproteins, DNA treatments, complex carbohydrates and viral particles are examples of biopharmaceuticals. Biopharmaceutical medications, whether intended for protein-based or human and veterinary use [124]. Although only a small portion of these items are made with biologically derived ingredients, the development of biopharmaceuticals is intimately related to the considerable breakthroughs in biotechnological approaches to protein chemistry.

In addition to CE's inherent advantages, including its high separation efficiency, quick separation time, small sample size, high resolution, and full automation capability [125], it also provides information on the isoforms, breakdown products, and contaminants of biopharmaceuticals as well as an analysis of their charge heterogeneity, opening up fascinating possibilities for effective investigations of intact peptides and proteins [126].

Mass spectrometry (MS) and sensitive detectors like laser-induced fluorescence (LIF) are two options for CE-based methods. It should be emphasized that for LIF detection, fluorescent moieties must be present inside the protein molecule [127]; certain proteins include aromatic amino acid residues with fluorescent characteristics, such as Trp, Tyr, and Phe, which may be recognized by the LIF detector directly [128].

6.4. Microfluidic Capillary Electrophoresis

Early in the 1990s, microchips, sometimes known as lab-on-a-chip technologies, were created. These tools for chemical sensing and analysis make use of the microfabrication processes that have advanced in the microelectronics sector [129]. The CEs chip reduction and online connection to other sensing systems. In these systems, a chip only a few square cm in size does the CE separation. In accordance with the sample injection mode, electrodes will be positioned in the various sample and background buffer reservoirs on the device [130]. After the sample is injected, the analytes are separated in the separation channels. The output of the microfluidic chip can then be coupled to a mass spectrometry detector, which is a common type of detection device [46].

Microfluidic devices have a number of benefits over benchtop instruments, including fast processing times, centralised sample volume control with practically no volume falloff, low reagent and sample volume utilisation, the ability to integrate various components onto a single system, and the ability to automate the necessary chemical analysis steps [131].

6.5. Affinity Ligands in Capillary Electrophoresis

Capillary electrophoresis is the application of a potential across a capillary to separate molecules according to their varied speeds of movement within the generated electrical field (CE) [132]. CE is known to offer a variety of advantages over traditional paper- or gel-based electrophoresis. These benefits include CE's capacity to deliver separations with high efficiency, quick turnaround times for analyses, and minimal sample requirements [133]. These properties have made CE a popular analytical technique for separating and analysing a range of substances in biological, pharmacological, and clinical samples for many years [134].

6.6. Capillary Electrophoresis in the Life Sciences

Capillary electromigration techniques have undergone extensive improvement and are now highly effective in numerous sectors [135]. High efficiencies can be attained with short separation durations and separation mostly depends on the charge-to-size ratio. In addition to capillary zone electrophoresis (CZE) for various applications, capillary isoelectric focusing (CIEF) and capillary gel electrophoresis (CGE) are effective methods for analysing biopolymers. When compared to liquid and gas chromatography [136], capillary electrophoresis (CE) was frequently less reproducible and robust, but over the past few decades, this has improved thanks to steady electroosmotic flows (EOF) and reliable injection. The creation and distribution of appropriate automated equipment were crucial to the advancement of CE [137].

For electrophoresis, which has long since been established, various development has been written. In 1937, Tiselius described protein separations using a U-tube, but the technique could only distinguish between two components, limiting its applicability. Thermal convection and diffusion resulted in a rather low separation efficiency. It should be noted that in 1930, Tiselius published his thesis on the application of electrophoresis to analysis [5]. Tiselius won the Nobel Prize for his contributions to separation science. A few new applications and developments were published over the ensuing years. Martin made a significant contribution by successfully employing displacement electrophoresis to separate glutamate, aspartate, acetate, and chloride (isotachophoresis, ITP) [138].

6.7. Capillary Electrophoresis in Forensic Analytical Chemistry

Two applications of CE technology that have expanded more quickly in recent years are forensic chemistry and toxicology. CE is a potent and relatively new analytical technique base forensic toxicologists that has proven useful for the examination of a wide range of various forensic materials [139]. Forensic scientists find CE to be an appealing methodology due to its resolution power, quick analysis time, exceptional separation, simplicity, versatility, low mass limits of detection (LODs), the economy of reagents, minimal sample requirements, and potential for direct sample injection without involved sample pre-treatments [140]. when it is necessary to separate the target solute from a sample matrix that may be quite complex and contain a lot of endogenous components [141].

It has been shown that capillary-electrophoresis (CE) techniques provide promising, efficient, and procedures that are affordable for separating a variety of compounds [49], including those used in forensic investigations. Commercially available CE tools that are dependable and automated have improved forensic science methodologies [142].

Toxicology, psychology, pathology, psychiatry, entomology, anthropology, odontology, and even pure forensic chemistry difficulties are all covered by the large, multidisciplinary field of applications known as "forensic sciences," which also includes diverse technology [143].

Forensic analytical chemistry, which is defined as a discipline applied to crime scene analysis and to law, is one of the areas of analytical chemistry where the nature of the sample and the use of the analytical chemical information play important roles in selecting and carrying out the right chemical-analysis technique. Characterization and quantification of chemical substances at trace levels are dealt with in this area of forensic sciences. These chemicals frequently include explosive compounds, gunshot residues, inks, and writing materials [139].

6.8. 2D Separation System of Capillary Electrophoresis

Due to advancements in micromechanics and microfluidics and the demand for more effective analytical tools for the analysis of complex (biological) samples, 2D techniques involving CE offer a potent tool for separation [144]. Strong separation strategies are required for the examination of complicated samples. In several sectors, 2D chromatographic separation methods (such LC-LC and GC-GC) are being used more. In contrast to LC and GC, electrophoretic separation methods exhibit distinct selectivity and extremely high separation efficiency [145]. As a result, 2D separation systems with at least one CE-based separation method present an intriguing option with the possibility of a high degree of orthogonality. However, specific coupling strategies are needed in CE due to the typically tiny volumes and powerful electrical fields [146].

In addition to 2D separation in a single capillary, an interface may perform online coupling of two separation processes. These systems are more adaptable than the single-capillary method and enable the blending of chromatographic and electrophoretic separation methods. For the first dimension [144], other CE modes have been utilized in addition to LC. CE has been the preferred approach for the second dimension. The following interface types have been used: hydrodynamic, flow gating, flow gating micro reactor, porous, tee-union, and dialysis [144].

6.9. Agrochemical Analysis

The effective use of CE for the determination of herbicides, insecticides, fungicides and acaricides in soil, water and foodstuffs is a key activity in the environmental monitoring of agrochemicals [147]. The uses of CE in the analysis of agrochemicals are comparable to those of CE in the study of pharmaceuticals. Impurity profiling, Assay, and chiral separations are some of them. Since many herbicides and pesticides, such as phenoxy acids, have an acidic character, they provide excellent separation candidates for CE analyses with elevated pH buffers [148]. Because they are highly basic in nature, quaternary ammonium agrochemicals like paraquat may be effectively evaluated at low pH levels, frequently with the addition of organic solvents in the buffer [30].

The development of a chiral CE method using cyclodextrin, which made it possible to resolve the chirality of 18 aromatic aminophosphonic acid enantiomers, is a recent example of chiral separation. The routine enantiopurity determination of aminophosphonic acids was effectively accomplished using this separation [30].

6.10. Analysis of Inorganic Ions in Environmental Samples

In the area of inorganic ions, CE is expanding quickly and is frequently contrasted with ion chromatography (IC) [149]. For the separation of industrial and environmental analytes such as chlorinated phenols, phenoxy acid herbicides, aromatic acids, acidic dyes, isomerized a- and b-hop acids, nitro aromatic and nitramine compounds, inorganic anions, linear benzene alkylsulfonates and charged polymers, CE method development and sensitivity enhancement strategies have been developed [150].

Among the pollutants whose separation and analysis are covered are polynuclear aromatic hydrocarbons, polychlorinated biphenyls and polychlorodibenzopdioxins, herbicides, phenols, nitroaromatic and other chemical warfare-related compounds, aromatic sulfonic acids, amines, phthalate esters, carboxylic acids, organometallic compounds, carbonyls, and dyes [151].

6.11. Separation of Vitamins and Minerals

One or two droplets of blood, air-dried blood samples collected on filter paper, or vitamin A and vitamin A binding protein can all be identified using CZE. This can be a helpful tool for determining newborn's and young children's vitamin A levels [152]. Vitamin C levels in urine, serum, and even fruit juices can be found using CE. By using CE, it is also able to locate and measure traces of iron in human serum. This approach may be useful for identifying iron-deficiency anaemia since it is sensitive enough to find iron in as low as one drop (10 μL) of serum [153].

Conclusion

Capillary electrophoresis is a technique which separates ions or charges on the basis of electrophoretic mobility and electroosmotic flow. Capillary electrophoresis has a short analytical time, high separation resolution, low waste generation, and have a broad range of applications. In this chapter, we discussed different parameters of capillary electrophoresis like its working, instrumentation, various factors different types of capillary electrophoresis, and various applications. This technique is further useful in agrochemical analysis, the study of vitamins and minerals, microfluidic chip electrophoresis, analysis of inorganic ions, forensic analytical chemistry, the study of chiral analysis, life science, and different bimolecular analysis. Capillary electrophoresis is best applied to the analysis of polar compounds, protein synthesis, ion separation and carbohydrates.

Conflict of Interest

The authors declare no conflict of interest.

References

[1] Rickard, E.C., M. Strohl, R. Nielsen, Correlation of electrophoretic mobilities from capillary electrophoresis with physicochemical properties of proteins and peptides, *Analytical biochemistry,* 197 (1991) 197-207.

[2] Shmidt, J.L. *Free flow electrophoresis in an alternating electric field along a semipermeable membrane,* Columbia University1990.

[3] Gordon, M.J., X. Huang, S.L. Pentoney Jr, R.N. Zare, Capillary electrophoresis, *Science,* 242 (1988) 224-228.

[4] Li, S.F.Y. *Capillary electrophoresis: principles, practice and applications,* Elsevier1992.

[5] Vesterberg, O. History of electrophoretic methods, *Journal of Chromatography A,* 480 (1989) 3-19.

[6] www.sciex.com/ce-features-and-benefits/history-of-ce.

[7] Nuccitelli, R., P. Nuccitelli, S. Ramlatchan, R. Sanger, P.J. Smith, Imaging the electric field associated with mouse and human skin wounds, *Wound repair and regeneration,* 16 (2008) 432-441.

[8] Bohrer, B.C., S.I. Merenbloom, S.L. Koeniger, A.E. Hilderbrand, D.E. Clemmer, Biomolecule analysis by ion mobility spectrometry, *Annual review of analytical chemistry* (Palo Alto, Calif.), 1 (2008) 293.

[9] Kenndler, E. A critical overview of non-aqueous capillary electrophoresis. Part II: separation efficiency and analysis time, *Journal of Chromatography A*, 1335 (2014) 31-41.

[10] Van Deemter, J., F. Zuiderweg, A. Klinkenberg, Longitudinal diffusion and resistance to mass transfer as causes of nonideality in chromatography, *Chemical Engineering Science,* 50 (1995) 3869-3882.

[11] Cohen, A., A. Paulus, B. Karger, High-performance capillary electrophoresis using open tubes and gels, *Chromatographia,* 24 (1987) 15-24.

[12] Swerdlow, H., R. Gesteland, Capillary gel electrophoresis for rapid, high resolution DNA sequencing, *Nucleic acids research,* 18 (1990) 1415-1419.

[13] Chakraborty, S., A.K. Srivastava, Generalized model for time periodic electroosmotic flows with overlapping electrical double layers, *Langmuir,* 23 (2007) 12421-12428.

[14] Jiang, W., J.N. Awasum, K. Irgum, Control of electroosmotic flow and wall interactions in capillary electrophosesis capillaries by photografted zwitterionic polymer surface layers, *Analytical chemistry,* 75 (2003) 2768-2774.

[15] Riskin, J., A. Khentov, *Electrocorrosion and protection of metals,* Elsevier 2019.

[16] Copper, C.L., K.W. Whitaker, Capillary Electrophoresis: Part II. Applications, *Journal of chemical education,* 75 (1998) 347.

[17] Foret, F., T.J. Thompson, P. Vouros, B.L. Karger, P. Gebauer, P. Bocek, Liquid sheath effects on the separation of proteins in capillary electrophoresis/ electrospray mass spectrometry, *Analytical Chemistry,* 66 (1994) 4450-4458.

[18] Skoog, D.A., F.J. Holler, S.R. Crouch, *Instrumental analysis,* Brooks/Cole, Cengage Learning Belmont2007.

[19] Tok, J., J. Lai, T. Leung, S.F.Y. Li, Selection of aptamers for signal transduction proteins by capillary electrophoresis, *Electrophoresis,* 31 (2010) 2055-2062.

[20] Whatley, H. Basic principles and modes of capillary electrophoresis, *Clinical and forensic applications of capillary electrophoresis,* Springer2001, pp. 21-58.

[21] Lele, M., S.M. Lele, J.R. Petersen, A. Mohammad, Capillary electrophoresis, *Clinical and Forensic Applications of Capillary Electrophoresis,* (2001) 3-19.

[22] Mack, S., I. Cruzado-Park, J. Chapman, C. Ratnayake, G. Vigh, A systematic study in CIEF: defining and optimizing experimental parameters critical to method reproducibility and robustness, *Electrophoresis,* 30 (2009) 4049-4058.

[23] Burgi, D.S., R.L. Chien, Optimization in sample stacking for high-performance capillary electrophoresis, *Analytical chemistry,* 63 (1991) 2042-2047.

[24] Taiwo, A.L. *Application of Capillary Electrophoresis for the Determination of Organoarsenicals in Poultry Farm Waters,* Thompson Rivers University, 2020.

[25] Hayes, M.A., A.G. Ewing, Electroosmotic flow control and monitoring with an applied radial voltage for capillary zone electrophoresis, *Analytical Chemistry,* 64 (1992) 512-516.

[26] Xu, X., L. Yu, G. Chen, Determination of flavonoids in Portulaca oleracea L. by capillary electrophoresis with electrochemical detection, *Journal of Pharmaceutical and Biomedical Analysis,* 41 (2006) 493-499.

[27] Lämmerhofer, M., F. Svec, J.M. Fréchet, W. Lindner, Separation of enantiomers by capillary electrochromatography, *TrAC Trends in Analytical Chemistry,* 19 (2000) 676-698.

[28] Lichtenberg, J., N.F. de Rooij, E. Verpoorte, A microchip electrophoresis system with integrated in-plane electrodes for contactless conductivity detection, *Electrophoresis,* 23 (2002) 3769-3780.

[29] Kubáň, P., F. Foret, G. Erny, Open source capillary electrophoresis, *Electrophoresis,* 40 (2019) 65-78.

[30] Altria, K., D. Elder, Overview of the status and applications of capillary electrophoresis to the analysis of small molecules, *Journal of Chromatography A,* 1023 (2004) 1-14.

[31] Ramautar, R., A. Demirci, G.J. de Jong, Capillary electrophoresis in metabolomics, *TrAC Trends in Analytical Chemistry,* 25 (2006) 455-466.

[32] Karlinsey, J.M. Sample introduction techniques for microchip electrophoresis: A review, *Analytica Chimica Acta,* 725 (2012) 1-13.

[33] Wang, J., G. Chen, A. Muck Jr, G.E. Collins, Electrophoretic microchip with dual-opposite injection for simultaneous measurements of anions and cations, *Electrophoresis,* 24 (2003) 3728-3734.

[34] Cao, P., M. Moini, A novel sheathless interface for capillary electrophoresis/ electrospray ionization mass spectrometry using an in-capillary electrode, *Journal of the American Society for Mass Spectrometry,* 8 (1997) 561-564.

[35] Brzyski, P., M. Gładecki, M. Rumińska, K. Pietrak, M. Kubiś, P. Łapka, Influence of hemp shives size on hygro-thermal and mechanical properties of a hemp-lime composite, *Materials,* 13 (2020) 5383.

[36] Breadmore, M.C. Electrokinetic and hydrodynamic injection: making the right choice for capillary electrophoresis, *Bioanalysis,* 1 (2009) 889-894.

[37] Blas, M., N. Delaunay, J.L. Rocca, Electrokinetic-based injection modes for separative microsystems, *Electrophoresis,* 29 (2008) 20-32.

[38] Chan, K.C., Y. Yu, S.H. Ng, H.K. Mak, Y.W. Yip, Y. van der Merwe, T. Ren, J.S. Yung, S. Biswas, X. Cao, Intracameral injection of a chemically cross-linked hydrogel to study chronic neurodegeneration in glaucoma, *Acta biomaterialia,* 94 (2019) 219-231.

[39] Paiva Jr, R.D., A.M.M. Mallmann, J.C. Santos, T.B.L. Kist, Polyimide removal, cleaving, and fusion splicing of cylindrical and square fused silica capillaries for new separation and detection layouts in capillary electrophoresis and chromatography, *Journal of Separation Science,* 44 (2021) 2438-2448.

[40] Righetti, P.G. *Capillary electrophoresis in analytical biotechnology: a balance of theory and practice,* CRC press1995.

[41] Wätzig, H., M. Degenhardt, A. Kunkel, Strategies for capillary electrophoresis: Method development and validation for pharmaceutical and biological applications, *Electrophoresis,* 19 (1998) 2695-2752.

[42] Belder, D., M. Ludwig, Surface modification in microchip electrophoresis, *Electrophoresis,* 24 (2003) 3595-3606.

[43] Carey, C.M. *A new phosphate ion selective electrode,* American University1992.

[44] Kappes, T., P.C. Hauser, Portable capillary electrophoresis instrument with potentiometric detection, *Analytical Communications,* 35 (1998) 325-329.

[45] Holland, L.A., N.P. Chetwyn, M.D. Perkins, S.M. Lunte, Capillary electrophoresis in pharmaceutical analysis, *Pharmaceutical research,* 14 (1997) 372-387.

[46] Kubáň, P., P.C. Hauser, Ten years of axial capacitively coupled contactless conductivity detection for CZE–a review, *Electrophoresis,* 30 (2009) 176-188.

[47] Ind, A., M.S. Bull, N.S.C. Ind, R. Med, V. Mycol, I. Analytik-Umweltanalytik, H.S. Ware, U. ODP, c *An International Journal for Rapid Communication in Chromatography, Electrophoresis, and Associated Techniques.*

[48] Corbera, M., M. Hidalgo, V. Salvado, P. Wieczorek, Determination of glyphosate and aminomethylphosphonic acid in natural water using the capillary electrophoresis combined with enrichment step, *Analytica chimica acta,* 540 (2005) 3-7.

[49] Lee, B.-L., C.-N. Ong, Comparative analysis of tea catechins and theaflavins by high-performance liquid chromatography and capillary electrophoresis, *Journal of chromatography A,* 881 (2000) 439-447.

[50] MacTaylor, C.E., A.G. Ewing, Critical review of recent developments in fluorescence detection for capillary electrophoresis, *Electrophoresis,* 18 (1997) 2279-2290.

[51] Götz, S., U. Karst, Recent developments in optical detection methods for microchip separations, *Analytical and Bioanalytical Chemistry,* 387 (2007) 183-192.

[52] Swinney, K., D.J. Bornhop, Detection in capillary electrophoresis, *Electrophoresis: An International Journal,* 21 (2000) 1239-1250.

[53] García-Cañas, V., C. Simo, M. Castro-Puyana, A. Cifuentes, Recent advances in the application of capillary electromigration methods for food analysis and Foodomics, *Electrophoresis,* 35 (2014) 147-169.

[54] Nouwairi, R.L., K.C. O'Connell, L.M. Gunnoe, J.P. Landers, Microchip electrophoresis for fluorescence-based measurement of polynucleic acids: recent developments, *Analytical Chemistry,* 93 (2020) 367-387.

[55] Gómez Laserna, O. *Development of in situ non destructive analytical methodologies for the conservation diagnosis of urban built heritage,* (2015).

[56] Revathi, C. Enzymatic and nonenzymatic electrochemical biosensors, *Fundamentals and sensing applications of 2D materials,* Elsevier2019, pp. 259-300.

[57] Gorecki, T. Chromatography, *Chromatographic analysis of the environment,* (2006) 133-176.

[58] Liu, C., E. Geva, M. Mauk, X. Qiu, W.R. Abrams, D. Malamud, K. Curtis, S.M. Owen, H.H. Bau, An isothermal amplification reactor with an integrated isolation

membrane for point-of-care detection of infectious diseases, *Analyst,* 136 (2011) 2069-2076.

[59] Lapainis, T.E. *Capillary electrophoretic technologies for single cell metabolomics,* University of Illinois at Urbana-Champaign2009.

[60] Baldwin, R.P. Recent advances in electrochemical detection in capillary electrophoresis, *Electrophoresis: An International Journal,* 21 (2000) 4017-4028.

[61] Guijt, R.M., C.J. Evenhuis, M. Macka, P.R. Haddad, Conductivity detection for conventional and miniaturised capillary electrophoresis systems, *Electrophoresis,* 25 (2004) 4032-4057.

[62] Simpson, D.C., R.D. Smith, Combining capillary electrophoresis with mass spectrometry for applications in proteomics, *Electrophoresis,* 26 (2005) 1291-1305.

[63] Hribar, J.A. Capillary Gradient Chromatofocusing-Mass Spectrometry*; a Sensitive Approach for Protein Analysis,* (2011).

[64] Baker, G.A., D.S. Moore, Progress in plasmonic engineering of surface-enhanced Raman-scattering substrates toward ultra-trace analysis, *Analytical and bioanalytical chemistry,* 382 (2005) 1751-1770.

[65] Tsai, C.H., P.H. Chan, C.H. Lin, T.C. Chang, C.T. Chia, A new approach for the detection of a nonfluorescent compound by CE-resonance Raman spectroscopy based on the sweeping-MEKC mode, *Electrophoresis,* 27 (2006) 4688-4693.

[66] Das, R.S., Y. Agrawal, Raman spectroscopy: recent advancements, techniques and applications, *Vibrational spectroscopy,* 57 (2011) 163-176.

[67] Theodoridis, G., H.G. Gika, I.D. Wilson, Mass spectrometry-based holistic analytical approaches for metabolite profiling in systems biology studies, *Mass spectrometry reviews,* 30 (2011) 884-906.

[68] Moerner, W., D.P. Fromm, Methods of single-molecule fluorescence spectroscopy and microscopy, *Review of Scientific instruments,* 74 (2003) 3597-3619.

[69] He, L., M.J. Natan, C.D. Keating, Surface-enhanced Raman scattering: A structure-specific detection method for capillary electrophoresis, *Analytical chemistry,* 72 (2000) 5348-5355.

[70] El Fellah, S., H.M.M. Siren, Recent developments in capillary electrophoresis of steroids and sterols, *Journal of Biomedical Research and Practice,* (2018).

[71] Wallingford, R.A., A.G. Ewing, Capillary zone electrophoresis with electrochemical detection, *Analytical chemistry,* 59 (1987) 1762-1766.

[72] Kuhr, W.G., C.A. Monnig, Capillary electrophoresis, *Analytical Chemistry,* 64 (1992) 389-407.

[73] Kubáň, P., P.C. Hauser, Fundamentals of electrochemical detection techniques for CE and MCE, *Electrophoresis,* 30 (2009) 3305-3314.

[74] Thibault, P., N.J. Dovichi, General instrumentation and detection systems including mass spectrometric interfaces, *Capillary electrophoresis: Theory and practice,* CRC Press LLC Boca Raton, FL1998, pp. 23-89.

[75] White, P.C., N.S. Lawrence, J. Davis, R.G. Compton, Electrochemical determination of thiols: a perspective, *Electroanalysis: An International Journal Devoted to Fundamental and Practical Aspects of Electroanalysis,* 14 (2002) 89-98.

[76] Chen, I.-C., C.-W. Whang, Capillary electrophoresis with amperometric detection using a porous cellulose acetate joint, *Journal of Chromatography A,* 644 (1993) 208-212.

[77] Huang, X., R.N. Zare, S. Sloss, A.G. Ewing, End-column detection for capillary zone electrophoresis, *Analytical chemistry,* 63 (1991) 189-192.

[78] Kappes, T., P.C. Hauser, Electrochemical detection methods in capillary electrophoresis and applications to inorganic species, *Journal of Chromatography A,* 834 (1999) 89-101.

[79] Colon, L.A., R. Dadoo, R.N. Zare, Determination of carbohydrates by capillary zone electrophoresis with amperometric detection at a copper microelectrode, *Analytical Chemistry,* 65 (1993) 476-481.

[80] Rodríguez-Plaza, P., R. González, M. Moreno-Arribas, M.C. Polo, G. Bravo, J.M. Martínez-Zapater, M. Martínez, A. Cifuentes, Combining microsatellite markers and capillary gel electrophoresis with laser-induced fluorescence to identify the grape (*Vitis vinifera*) variety of musts, *European Food Research and Technology,* 223 (2006) 625-631.

[81] Berezovski, M., S.N. Krylov, Nonequilibrium capillary electrophoresis of equilibrium mixtures− a single experiment reveals equilibrium and kinetic parameters of protein− DNA interactions, *Journal of the American Chemical Society,* 124 (2002) 13674-13675.

[82] Hughes, M.P. Strategies for dielectrophoretic separation in laboratory-on-a-chip systems, *Electrophoresis,* 23 (2002) 2569-2582.

[83] Drossman, H., J.A. Luckey, A.J. Kostichka, J. D'Cunha, L.M. Smith, High-speed separations of DNA sequencing reactions by capillary electrophoresis, *Analytical Chemistry,* 62 (1990) 900-903.

[84] Upcroft, P., J.A. Upcroft, Comparison of properties of agarose for electrophoresis of DNA, *Journal of Chromatography B: Biomedical Sciences and Applications,* 618 (1993) 79-93.

[85] Maurer, H.R. Disc electrophoresis and related techniques of polyacrylamide gel electrophoresis, *Disc Electrophoresis and Related Techniques of Polyacrylamide Gel Electrophoresis,* de Gruyter 2011.

[86] Corni, I., M.P. Ryan, A.R. Boccaccini, Electrophoretic deposition: From traditional ceramics to nanotechnology, *Journal of the European Ceramic Society,* 28 (2008) 1353-1367.

[87] Silva, M., D. Pérez-Quintanilla, S. Morante-Zarcero, I. Sierra, M.L. Marina, Z. Aturki, S. Fanali, Ordered mesoporous silica functionalized with β-cyclodextrin derivative for stereoisomer separation of flavanones and flavanone glycosides by nano-liquid chromatography and capillary electrochromatography, *Journal of Chromatography A,* 1490 (2017) 166-176.

[88] Bartle, K.D., P. Myers, Factors Influencing Performance in Capillary Electrochromatography, *Electrokinetic Phenomena: Principles and Applications in Analytical Chemistry and Microchip Technology,* (2003) 167.

[89] Virtanen, V., G. Bordin, A.-R. Rodriguez, Separation of metallothionein isoforms with capillary zone electrophoresis using an uncoated capillary column effects of

pH, temperature, voltage, buffer concentration and buffer composition, *Journal of Chromatography A,* 734 (1996) 391-400.
[90] Nováková, L., H. Vlčková, A review of current trends and advances in modern bio-analytical methods: chromatography and sample preparation, *Analytica chimica acta,* 656 (2009) 8-35.
[91] Hadley, M., M. Gilges, J. Senior, A. Shah, P. Camilleri, Capillary electrophoresis in the pharmaceutical industry: applications in discovery and chemical development, *Journal of Chromatography B: Biomedical Sciences and Applications,* 745 (2000) 177-188.
[92] Kalyankar, T., P. Kulkarni, S. Wadher, S. Pekamwar, Applications of micellar liquid chromatography in bioanalysis: a review, *Journal of Applied Pharmaceutical Science,* 4 (2014) 128-134.
[93] Poole, C.F., S.K. Poole, Column selectivity from the perspective of the solvation parameter model, *Journal of Chromatography A,* 965 (2002) 263-299.
[94] Shamsi, S.A. Micellar electrokinetic chromatography− mass spectrometry using a polymerized chiral surfactant, *Analytical chemistry,* 73 (2001) 5103-5108.
[95] Burlatsky, S.F., V.V. Atrazhev, D.V. Dmitriev, V.I. Sultanov, E.N. Timokhina, E.A. Ugolkova, S. Tulyani, A. Vincitore, Surface tension model for surfactant solutions at the critical micelle concentration, *Journal of colloid and interface science,* 393 (2013) 151-160.
[96] Terabe, S., N. Matsubara, Y. Ishihama, Y. Okada, Microemulsion electrokinetic chromatography: comparison with micellar electrokinetic chromatography, *Journal of Chromatography A,* 608 (1992) 23-29.
[97] Moghadam, B.Y., K.T. Connelly, J.D. Posner, Isotachophoretic preconcenetration on paper-based microfluidic devices, *Analytical chemistry,* 86 (2014) 5829-5837.
[98] Peterson, Z.D., C.R. Bowerbank, D.C. Collins, S.W. Graves, M.L. Lee, Advantages and limitations of coupling isotachophoresis and comprehensive isotachophoresis–capillary electrophoresis to time-of-flight mass spectrometry, *Journal of Chromatography A,* 992 (2003) 169-179.
[99] Jorgenson, J.W., K.D. Lukacs, Zone electrophoresis in open-tubular glass capillaries, *Analytical Chemistry,* 53 (1981) 1298-1302.
[100] Ahuja, S. *Chromatography and separation science,* Academic Press2003.
[101] Vesterberg, O., H. Svensson, Isoelectric fractionation, analysis, and characterization of ampholytes in natural pH gradients, *Acta chem. scand,* 20 (1966) 820-834.
[102] Righetti, P.G. Determination of the isoelectric point of proteins by capillary isoelectric focusing, *Journal of chromatography A,* 1037 (2004) 491-499.
[103] Mazzeo, J.R., I.S. Krull, Capillary isoelectric focusing of proteins in uncoated fused silica capillaries using polymeric additives, *Analytical chemistry,* 63 (1991) 2852-2857.
[104] Jochheim, C., S. Novick, A. Balland, J. Mahan-Boyce, W.-C. Wang, A. Goetze, W. Gombotz, Separation of Enbrel®(rhuTNFR: Fc) isoforms by capillary isoelectric focusing, CE in Biotechnology: *Practical Applications for Protein and Peptide Analyses,* Springer2001, pp. 59-65.
[105] Pandey, R. *Techniques in Life Science,* BFC Publications2021.

[106] Lewis, A., A. Cranny, N. Harris, N.G. Green, J.A. Wharton, R. Wood, K.R. Stokes, Review on the development of truly portable and in-situ capillary electrophoresis systems, *Measurement Science and Technology,* 24 (2013) 042001.

[107] Weinberger, R. *Practical capillary electrophoresis,* Elsevier2000.

[108] Maizel Jr, J.V. Polyacrylamide gel electrophoresis of viral proteins, *Methods in virology,* Elsevier1971, pp. 179-246.

[109] Lobaskin, V., B. Dünweg, C. Holm, Electrophoretic mobility of a charged colloidal particle: a computer simulation study, *Journal of Physics: Condensed Matter,* 16 (2004) S4063.

[110] Vasiliev, A., A. Sukhanovskii, P. Frick, A. Budnikov, V. Fomichev, M. Bolshukhin, R. Romanov, High Rayleigh number convection in a cubic cell with adiabatic sidewalls, *International Journal of Heat and Mass Transfer,* 102 (2016) 201-212.

[111] Swinney, K., D.J. Bornhop, Quantification and evaluation of Joule heating in on-chip capillary electrophoresis, *Electrophoresis,* 23 (2002) 613-620.

[112] Zhmakin, A. Heat conduction beyond the Fourier law, *Technical Physics,* 66 (2021) 1-22.

[113] Pantelic, J., S. Liu, L. Pistore, D. Licina, M. Vannucci, S. Sadrizadeh, A. Ghahramani, B. Gilligan, E. Sternberg, K. Kampschroer, Personal CO2 cloud: laboratory measurements of metabolic CO2 inhalation zone concentration and dispersion in a typical office desk setting, *Journal of exposure science & environmental epidemiology,* 30 (2020) 328-337.

[114] Grushka, E., R. McCormick, J. Kirkland, Effect of temperature gradients on the efficiency of capillary zone electrophoresis separations, *Analytical Chemistry,* 61 (1989) 241-246.

[115] Diamantopoulos, E., W. Durner, T. Harter, Prediction of capillary air-liquid interfacial area vs. saturation function from relationship between capillary pressure and water saturation, *Advances in Water Resources,* 97 (2016) 219-223.

[116] Pappenheimer, J.R. Passage of molecules through capillary walls, *Physiological Reviews,* 33 (1953) 387-423.

[117] Řezanka, P., K. Navrátilová, M. Řezanka, V. Král, D. Sýkora, Application of cyclodextrins in chiral capillary electrophoresis, *Electrophoresis,* 35 (2014) 2701-2721.

[118] Chankvetadze, B. *Capillary electrophoresis in chiral analysis,* John Wiley & Sons1997.

[119] Herrero, M., C. Simó, V. García-Cañas, S. Fanali, A. Cifuentes, Chiral capillary electrophoresis in food analysis, *Electrophoresis,* 31 (2010) 2106-2114.

[120] Saz, J,. M. Marina, Recent advances on the use of cyclodextrins in the chiral analysis of drugs by capillary electrophoresis, *Journal of Chromatography A,* 1467 (2016) 79-94.

[121] Bernardo-Bermejo, S., E. Sánchez-López, M. Castro-Puyana, M.L. Marina, Chiral capillary electrophoresis, *TrAC Trends in Analytical Chemistry,* 124 (2020) 115807.

[122] Kristoff, C.J., L. Bwanali, L.M. Veltri, G.P. Gautam, P.K. Rutto, E.O. Newton, L.A. Holland, Challenging bioanalyses with capillary electrophoresis, *Analytical chemistry,* 92 (2019) 49-66.

[123] Galievsky, V.A., A.S. Stasheuski, S.N. Krylov, Capillary electrophoresis for quantitative studies of biomolecular interactions, *Analytical Chemistry*, 87 (2015) 157-171.

[124] Twyman, R.M., S. Schillberg, R. Fischer, Transgenic plants in the biopharmaceutical market, *Expert opinion on emerging drugs,* 10 (2005) 185-218.

[125] Meibohm, B. *The role of pharmacokinetics and pharmacodynamics in the development of biotech drugs, Pharmacokinetics and Pharmacodynamics of Biotech Drugs: Principles and Case Studies in Drug Development.* Weinheim, Germany: Wiley-VCH, (2006) 3-16.

[126] Deshaies, R.J. Multispecific drugs herald a new era of biopharmaceutical innovation, *Nature,* 580 (2020) 329-338.

[127] Moschou, E.A., B.V. Sharma, S.K. Deo, S. Daunert, Fluorescence glucose detection: advances toward the ideal in vivo biosensor, *Journal of fluorescence,* 14 (2004) 535-547.

[128] Tamizi, E., A. Jouyban, The potential of the capillary electrophoresis techniques for quality control of biopharmaceuticals—A review, *Electrophoresis,* 36 (2015) 831-858.

[129] Becker, H., C. Gärtner, Polymer microfabrication methods for microfluidic analytical applications, *Electrophoresis: An International Journal,* 21 (2000) 12-26.

[130] Paul, P. *Capillary electrophoresis as a low-cost tool for pharmaceutical analysis,* (2018).

[131] Ramos-Payán, M., J.A. Ocaña-Gonzalez, R.M. Fernández-Torres, A. Llobera, M.Á. Bello-López, Recent trends in capillary electrophoresis for complex samples analysis: a review, *Electrophoresis,* 39 (2018) 111-125.

[132] Li, S., Y. Wu, Capillary electrophoresis, *Electrophoresis,* (2000) 1176-1187.

[133] Ghosal, S. Electrokinetic flow and dispersion in capillary electrophoresis, *Annual Review of Fluid Mechanics,* 38 (2006) 309-338.

[134] Zhang, C., A.G. Woolfork, K. Suh, S. Ovbude, C. Bi, M. Elzoeiry, D.S. Hage, Clinical and pharmaceutical applications of affinity ligands in capillary electrophoresis: A review, *Journal of pharmaceutical and biomedical analysis,* 177 (2020) 112882.

[135] Altria, K.D., Overview of capillary electrophoresis and capillary electrochromatography, *Journal of Chromatography A,* 856 (1999) 443-463.

[136] Jorgenson, J.W., K.D. Lukacs, Capillary zone electrophoresis, *Science,* 222 (1983) 266-272.

[137] Ohno, K.i., K. Tachikawa, A. Manz, Microfluidics: applications for analytical purposes in chemistry and biochemistry, *Electrophoresis,* 29 (2008) 4443-4453.

[138] Toraño, J.S., R. Ramautar, G. de Jong, Advances in capillary electrophoresis for the life sciences, *Journal of Chromatography B,* 1118 (2019) 116-136.

[139] Cruces-Blanco, C., L. Gámiz–Gracia, A. Garcia-Campana, Applications of capillary electrophoresis in forensic analytical chemistry, *TrAC Trends in Analytical Chemistry,* 26 (2007) 215-226.

[140] Favretto, D., J.P. Pascali, F. Tagliaro, New challenges and innovation in forensic toxicology: focus on the "New Psychoactive Substances", *Journal of chromatography A,* 1287 (2013) 84-95.

[141] Marquet, P., G. Lachatre, Liquid chromatography–mass spectrometry: potential in forensic and clinical toxicology, *Journal of Chromatography B: Biomedical Sciences and Applications,* 733 (1999) 93-118.

[142] Guzman, N.A., T.M. Phillips, Immunoaffinity capillary electrophoresis: a new versatile tool for determining protein biomarkers in inflammatory processes, *Electrophoresis,* 32 (2011) 1565-1578.

[143] James, S.H., J.J. Nordby, *Forensic science: an introduction to scientific and investigative techniques,* CRC press2002.

[144] Kohl, F.J., L. Sánchez-Hernández, C. Neusüß, Capillary electrophoresis in two-dimensional separation systems: Techniques and applications, *Electropho-resis,* 36 (2015) 144-158.

[145] Cao, J.-L., J.-C. Wei, M.-W. Chen, H.-X. Su, J.-B. Wan, Y.-T. Wang, P. Li, Application of two-dimensional chromatography in the analysis of Chinese herbal medicines, *Journal of Chromatography A,* 1371 (2014) 1-14.

[146] Voeten, R.L., I.K. Ventouri, R. Haselberg, G.W. Somsen, Capillary electrophoresis: trends and recent advances, *Analytical chemistry,* 90 (2018) 1464-1481.

[147] van der Hoff, G.R., P. van Zoonen, Trace analysis of pesticides by gas chromatography, *Journal of Chromatography A,* 843 (1999) 301-322.

[148] Aydoğan, C. Nanoscale separations based on LC and CE for food analysis: A review, *TrAC Trends in Analytical Chemistry,* 121 (2019) 115693.

[149] Haddad, P.R., P. Doble, M. Macka, Developments in sample preparation and separation techniques for the determination of inorganic ions by ion chromatography and capillary electrophoresis, *Journal of Chromatography A,* 856 (1999) 145-177.

[150] McLaughlin, G., A. Weston, K. Hauffe, Capillary electrophoresis methods development and sensitivity enhancement strategies for the separation of industrial and environmental chemicals, *Journal of Chromatography A,* 744 (1996) 123-134.

[151] Fukushi, K., S. Takeda, K. Chayama, S.-I. Wakida, Application of capillary electrophoresis to the analysis of inorganic ions in environmental samples, *Journal of Chromatography A,* 834 (1999) 349-362.

[152] Ma, Y., Z. Wu, H.C. Furr, C. Lammi-Keefe, N.E. Craft, Fast minimicroassay of serum retinol (vitamin A) by capillary zone electrophoresis with laser-excited fluorescence detection, *Journal of Chromatography B: Biomedical Sciences and Applications,* 616 (1993) 31-37.

[153] Shi, H., Y. Ma, J.H. Humphrey, N.E. Craft, Determination of vitamin A in dried human blood spots by high-performance capillary electrophoresis with laser-excited fluorescence detection, *Journal of Chromatography B: Biomedical Sciences and Applications,* 665 (1995) 89-96.

Chapter 6

High-Performance Thin Layer Chromatography and Its Hyphenated Techniques

Nikita Sharma
Debashree Debasish Das
and Pooja A. Chawla*
Department of Pharmaceutical Analysis, ISF College of Pharmacy, Ghal Kalan, Moga, Punjab, India

Abstract

High-performance thin-layer chromatography (HPTLC) is a sophisticated analytical instrument that is an automated version of thin-layer chromatography. The advancements like automation, optimization, the lower amount of sample and solvent system consumption and hyphenation, etc. make this analytical tool an essential and cost-effective analytical technique mainly used to gather chromatographic information and separate complex organic and biomolecular samples. This chapter contains the history of the advancements in TLC which led to the HPTLC method's development. This chapter is an overview of the instrumentation and the detailed working process of HPTLC with its advancements like automation and different types of hyphenation techniques with LC, MS, Raman spectrum, and FTIR.

Keywords: HPTLC, hyphenation, MS, Raman spectrum, FTIR

---

* Corresponding Author's Email: pvchawla@gmail.com.

In: Modern Approaches in Fluid Chromatography
Editors: Rohit Bhatia and Bhupinder Kumar
ISBN: 979-8-88697-856-8

1. Introduction

To enhance the understanding and selectivity of a process or technique, chemical analysts mainly emphasize new findings and claims. After the advancements in the methods are reserved as still intentionally as the information is compared for an extended-ages, analytical chemists are also anxious about the adjustments and advances made in the novel instruments. Also, the devices that are generally being advanced and executed currently include the tools like rotational, optical, absorption, colorimetric and scattering spectroscopy, mass spectrometry, chromatography, electro-chemicals, acoustics, laser, chemical imaging, light-induced fluorescence, light scattering also, the vibrational and rotational tools are included. All phases of the drug research and drug advancement process frequently involve examining pharmaceutical substances [1]. Such analytical methods produce more exact and accurate information, helping to monitor the market after the drug is released and having knowledge of the medication that is being researched and developed. Analysts always strive to make the already-used methods more reliable to meet the need for improved chemical assessments [2]. The pharmaceutical and the natural compounds and the newly discovered medication analysis are done in all the drug discovery phases and the advancement methods. We have discussed chromatography, which includes different types of traditional separation techniques as well as modified or advanced separation techniques [3].

The chromatography principle includes a separation process where the constituents or particles distributed over the mobile phase get separated faster than the particles or components distributed over the stationary phase. These separations are based on adsorption, partition, and the variances in the molecular mass of the solutes [4].

During the advancements in the separation, specific processes might occur during the separation process. The method includes that the sample particles are eluting first according to their affinity towards the stationary phase. Thus, the higher the affinity higher will the retention time of the particle and it will elute later compared to the particles with higher affinity towards the stationary phase. The other process is that to keep the separation that has been attained, the band's desire for expansion or distribution is restrained as they travel away from each other [4].

These methods process gives three types of elements that includes:

- Stationary phase

- Mobile phase
- Separated molecules

The stationary phase may be either solid or liquid, layered over a solid surface. At the same time, the mobile phase is the phase that is either liquid or gas. If the mobile phase is liquid, it is said to be liquid chromatography, usually for non-volatile compounds and solid compounds and if the mobile phase is a gaseous phase, it is said to be gas chromatography used for the separation of macromolecules. This method is used to get the desired result in a required time interval where an appropriate partition can be retained [5].

2. History

Meinhard and Hall, in 1949, labeled a term as surface chromatography, where the separation of the inorganic ions to provide an inflexible layer utilized a starch binder. Some developments were made in 1951 by Kirchner *et al.*, who employed the standard ascending method with the stationary phase where adsorption takes place made of silicic acid. Then again, Reitsema in 1954 came with the advancement by using broader plates with the help of which the separation was easy to conduct within one run. As Stahl from the year 1956 bought the sequence of papers that presented the analytical method known as thin layer chromatography, the silica gel from this time became the stationary phase also, the plaster of paris was used as a binder and the wide use of TLC since then came into existence [6]. The advantages of TLC, like its adaptability and operation in a wide range spectrum for the separation process, have also been brought to light because of its ability to separate quickly and cost-effectively. Although, with the help of the spectro densitometric scanners, which allow the measurement of the analytes straight from the TLC plate. Earlier, these measurements of the peak area were done physically; now, due to the automation of the methods, it is measured automatically [1].

In 1973 Halpaap was the one who first used small-sized silica particles while the preparation of the plates of TLC by comparing the plate heights as well as particle size and thus, by observing the result, it was then found to be a better method as compared to TLC for the separation process. When it became clear that HPTLC had given TLC a fresh perspective, its advertising name was switched from nano-TLC plates to HPTLC plates. Also, this method was found to be a method that uses less solvent, less analysis time, and more precision. Zlatkis and Kaiser's paper was the first significant paper on HPTLC (1977). In 1980, Halpaap *et al.* published reversed phase HPTLC. For several lipid-loving dyes, Halpaap and Ripphahn compared their outcomes using the improved 5.5-cm plates of HPTLC to the traditional TLC [1, 3].

3. Advancements

Modern pharmaceutical analysis is mainly dominated by costlier instrumental analysis. As a result, many analysts concentrate on creating add-ons as their applications, findings, and new analytical techniques that boost the method's precision and responsiveness. [2]. This advancement includes thin-layer chromatography, the most widely used method of separation in every field of the pharmaceutical industry. One of the important identification tests in most pharmacopeial monographs involves thin-layer chromatography. Industries often use pharmacopeia as a reference for adhering to the quality of the sample specifications with current good manufacturing practices (cGMPs). High-performance thin layer chromatography (HPTLC), a modified TLC technique, is a reliable, easy-to-use, quick, and effective technique to quantify the number of compounds by chemical analysis [2]. The advancements in the TLC that lead to the development of HPTLC are mentioned in Figure 1 [7].

TLC	Features	HPTLC
Manual	Technique	Instrumental
Lab made/pre-coated	Layer	Pre-coated
Circular (2-4 mm diameter)	Shape of sample	Rectangular (6 mm L x 1 mm)
Uncontrolled solvent dependent	Size of sample	Controlled solvent independent
2-10%	Volume precision	0.5%-2%
Non-compliant	Good lab practice	Compliant
1 μL-10 μL	Volume range	0.1 μL -500 μL
No	Method storage	Yes
No	Validation	Yes
Capillary/Pipette	Sample holder	Syringe
254 nm or 366 nm	Wavelength range	190 or 800nm
No	Quantitative analysis	Yes
No	Spectrum analysis	Yes
By analyst	Analysis judgement	By machine

Figure 1. Advancements in TLC for the development of HPTLC.

4. HPTLC

High-performance thin-layer chromatography (HPTLC), a planar chromatographic technique significantly more advanced than classic thin-layer chromatography that uses HPTLC plates, created as a result of the evolution of traditional thin-layer chromatography (TLC). Furthermore, the separation that arises from the action taking place between the sample's contents with a planar stationary phase, a mobile phase, may it be a liquid phase, or a gaseous phase formed in the chromatographic chamber and thus, HPTLC and TLC share the similar fundamental principles in between them. Additionally, they have benefits in common such as visible findings, simultaneous sample evaluation, an evaluation that can be done in single plate, quick results, adaptability, and the potential for numerous detections. Some technical differences are there between HPTLC and TLC, concerning the plate, solvent consumption, duration of development and sensitivity [8]. There are also a few variances among the HPTLC and TLC regarding the plate, usage of the mobile phase, time taken for the development and also the efficiency of the technique.

Apart from these distinctions and analogies, HPTLC is a novel idea, with the main emphasis on replicating and separating ability. It employs procedures that are clearly described, have standardized and optimized parameters, undergo a verification, i.e., validation procedure, and produces accurate analytical findings with high intra- and inter-laboratory repeatability [8].

HPTLC equipment ranges from basic to advanced, enabling the production of detectable digital pictures and more thorough use of the data [8]. Precision control over sample application, chromatographic development, derivatization, identification, and determination is made possible by the optimized instrumentation in conjunction with high rates of an automated technology. The distinct quality of HPTLC is the capability to parallelly spot and develop and detect numerous separations over a single plate. Naturally, the loading sample of amount 2 to 5 μL with about 18 to 36 samples able to be detected in a single container with a time duration of 3 to 20 minutes for covering the path of approximately 2 to 7 cm. Although HPTLC exhibits a more extensive detection range than HPLC, it can be resolved by the bioautography technique or post-chromatographic conversion of the existing sample state. It has been made compulsory to use the pressured flow of the mobile phase rather than capillary flow due to the noticeably more extended processing period required in the HPTLC chromatography plates [9]. Additionally, HPTLC is referred to as GMP-friendly. Detecting various samples simultaneously provides adaptability toward TLC, and HPTLC enables the samples extracted from natural plants to be identified or detected in a single analysis [8].

For the fast detection of numerous samples, the high-throughput assessment method is utilized with HPTLC. The demand for supporting many pharmacological targets developing from molecular biology, human genetics, and functional genomics has accelerated the development in this field. Additionally, financial pressure to shorten the requirement for novel medication candidates to come into the market and support the analysis of huge chemical collections resulting from concurrent chemistry have been a driving force for the progress [2].

HPTLC is a widely utilized analytical technique where the use of modern equipment like video scanners, densitometers, and the latest chromatographic development chambers with more efficient elution techniques, highly efficient adsorbents with a preferred size of particles also the ability to combine with different instrumental methods with the establishment of computer programs are included for the method enhancement. By utilizing the readily accessible commercialized tools, different phases of HPTLC method development were entirely automated and the total procedure could well be managed using the software that complies with drug regulatory authorities' standards. The methods based on the HPTLC technique are being used and taken as an upright alternative also because of being discovered as a crucial technique in the repetitive analysis [10].

The steps involved in the method of HPTLC are shown in Figure 2.

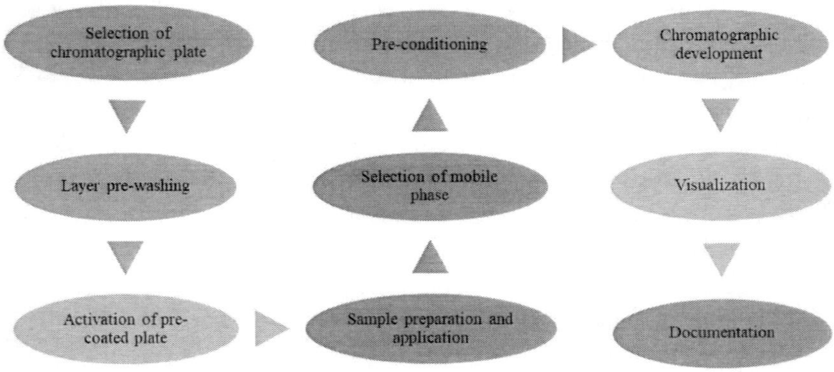

Figure 2. Steps involved in HPTLC.

4.1. Selection of Chromatographic Plate/Preparation of Plate

4.1.1. Precoated Layers

TLC plates can be prepared in any laboratory with the help of the appropriate apparatus. Although these layers do not get attached easily over the glass plate that acts as a support.

With the use of precoated plates made of heavy polymers in small amounts, the limitation of the layer of the plate that is made in laboratories can be resolved. Precoated layers are activated before their use, repeatable, somewhat attrition resistant and have highly consistent layer thickness with aluminum, glass or polyester support. The aluminum foil plates are easier to handle, transfer and mail as they are cheaper to purchase and may be sliced. The glass plates achieve the most significant outcomes. Coatings with the fluorescent indication F 254 are typically utilized. This makes it possible to quickly, easily, and non-destructively see samples in a UV chamber. The regularly used TLC plate size is 20×20 cm and in HPTLC 20 × 10 cm or 10 × 10 cm [10].

4.2. Layer Prewashing

Places are usually lifted by holding at the upper side to minimize contamination. Until the chromatography develops impurity fronts from contaminated plates, which are often utilized without any preparation. Layers are frequently prewashed in a 20 x 10 cm twin-trough chamber by using 20 mL methanol (usually, methanol is used as a solvent for prewashing; although, a combination of methanol and ethyl acetate as well as the technique's mobile phase could also be employed) per trough (TTC, twin trough chambers). Within every TTC trough, a maximum of up to two 20 x 10 cm or four 10 x 10 cm plates could be produced one after the other. Balance the plate by the atmospheric condition with a laboratory's specific temperature and humidity by guarding the plate against pollution [11].

4.3. Application of Samples and Standards

In HPTLC, compact, homogenous sample bands or streaks of specified length are often administered using an automated applicator such as the CAMAG

Linomat 4 having 100 mL syringe, CAMAG Automatic TLC Sampler (Wilmington, NC).

Spray-on band application using all these devices is the method that is most commonly utilized for HPTLC drug analysis as one larger range of quantities may be sprayed with tighter areas where concentrated sample are used, the scanning densitometry obtains high-resolution separations, consistent and exact quantitative data.

For preparative separations and qualitative and quantitative analysis, the Linomat 5 allows semi-automatic sample application. For everyday usage with moderate sample output, the device is acceptable. In Linomat, for the change of sample, an operator's involvement is necessary but not required in the Automatic TLC Sampler (ATS 4).

The HPTLC lab's output dramatically depends upon the automatic application of samples. The Automatic TLC Sampler satisfies each criterion for an instrument intended for this task, including accuracy, durability in everyday usage, and ease of use [12].

4.4. Selection of Mobile Phase

Depending upon the stationary phase's sorbent and the analyte's chemical and physical characteristics, the mobile phase is chosen [11]. Resolution of the desired peaks should occur between Rf values of 0.15 and 0.85. Eluent intensity, a factor associated with the polarity of the mobile phase constituents, determines how powerfully the mobile phase gets eluted. The chemical elution depends upon their polarity. If the chemical is nonpolar, it will elute faster (or spend a shorter period in the stationary phase); if the chemical is polar, it will elute slower (or remains for more time in the stationary phase). Compared to TLC, HPTLC requires a minimal amount of mobile phase [13, 14]. Various solvents are used for separation. Some of them are mentioned in Table 1 with their elution power.

4.5. Preconditioning (Chamber Saturation)

The separation pattern is significantly impacted by chamber saturation. The period needed for saturation relies on the mobile phase. When the plates are exposed to the chamber throughout the development process, the solvents vanish due to evaporation, resulting in bigger Rf values. If the saturation of the TLC chamber is done before the development process, the solvent's vapors get equally dispersed all over the chamber. Smaller Rf values are produced as a result of the plates rapidly getting preloaded by the solvent vapors after being placed inside the saturated chamber. As a result, a minimal number of solvents are needed to cover a given distance. However, based on the polarity, there are specific circumstances when chamber saturation is necessary or not. The filter paper helps in the chamber's saturation and distributes the vapor equally over the room and, thus, requires less amount of mobile phase to cover the distance for the development [15, 16].

Table 1. Some common solvents [13]

S. No.	Solvent	Elution Power
1.	N- Pentane	0.00
2.	Hexane	0.01
3.	Cyclohexane	0.04
4.	Carbon tetrachloride	0.18
5.	Toluene	0.29
6.	Chloroform	0.40
7.	Methylene chloride	0.42
8.	Tetrahydrofuran	0.45
9.	Acetone	0.56
10.	Ethyl acetate	0.58
11.	Aniline	0.62
12.	Acetonitrile	0.65
13.	Ethanol	0.88
14.	Methanol	0.95

4.6. Chromatographic Development and Drying

HPTLC plates are usually grown in horizontal development chambers, twin trough chambers, or flat bottom chambers, i.e., the different developmental techniques used are - ascending, descending and horizontal developmental methods [15]. The saturated twin trough chambers (TTC) where filter papers are used provide better repeatability. A proper amount of mobile phase is required in advance for the plate development process in a saturated TTC. A piece of filter paper cut in the proper size should be placed in the TTC's back trough and attached to the wall behind it. The plate is tilted up to about 45°C to avoid uneven solvent distribution between both the troughs. Keep the chamber stably in place, then let the chamber equilibrize for about 20 minutes by placing the lid over it. The required distance for the further determination of the sample is marked with a pencil over the right side of the plate i.e., 70 mm above the lower end. Put the plate into the front trough with the coating/layer and filter paper facing each other and the plate's backside lying towards the TTC's front wall after sliding the lid off to the side. The plate should be taken out of the chamber and dried for five minutes in a flow of cold air while being held vertically in the direction of the chromatography [17].

4.7. Visualization at UV 254 nm

Phosphorescence quenching is also termed F254. It was found to be comparatively short and lasts for more than ten seconds. Green fluorescence is produced whenever the F254 fluorescent indicator is stimulated by UV light at a wavelength of 254. A dark violet patch on a green backdrop is where the components zones are placed. Compounds that absorb light at 254 nm lower the emission on the layer.

4.8. Visualization at UV 366 nm

Fluorescence quenching is the proper term for F366. In this case, removing the origin of excitation leads to the disappearance of the fluorescence. Anthraglycosides, coumarins, flavonoids, phenol carboxylic acids, and several alkaloid types exhibit quenching (Rauwolfia, Ipecacuanha alkaloids).

4.9. Visualization at White Light

In this zone, the natural color in daylight can be used to identify its distinct constituents (White light) [13].

4.10. Derivatization

Derivatization is a procedure that mainly alters an analyte's features that allow chromatographic separations. This process can be carried out by submerging the plates inside an appropriate reagent or by sprinkling or spraying the plates with the reagent. Immersion is the recommended derivatization technique for better reliability [18].

4.11. Key Features and Separation Efficiency

Even though the most widely used sorbent known as the stationary phase, is the silica gel, reverse phases, bonded phases, alumina, Kieselguhr, etc. are some of the various other adsorbents that are also utilized as a separate source. The solvents are also used as a mobile phase. The vapor saturation, pH, and humidity of the developing chamber are all critical factors of the TLC/HPTLC process. For qualitative, semi-quantitative, and quantitative analysis, TLC and HPTLC can indeed be employed. It could also be used to identify industrial fractions following chromatographic separations and to identify natural substances and complicated mixes using an "HPTLC fingerprint." For contaminant analysis, assay, comparison with related samples, testing of unidentified samples, or a numerous sample set, TLC/HPTLC is typically used in labs. For routine analysis, TLC/HPTLC is a helpful technique in the fields of quality control, analytical R&D, performance monitoring, and laboratories related to the environment. In TLC/HPTLC, the development of a chromatogram can be carried out either vertically or horizontally following the plate. The movement of the mobile phase also occurs in a linear, circular, or anti-circular motion to accomplish. Dimension development is also possible where only one mobile phase is used, also known as an isocratic elution, or with different mobile phases in the same direction, known as a gradient elution system. For high-resolution separations, two-dimension chromatography is highly helpful when a separate mobile phase is used in each order. Using a two-dimension method, the chromatographic stability of samples may be examined [19].

4.12. Automation

Automation is a development that will enable an analytical technique to complete all steps, such as sampling, sample preparation and dissolution, interference removal, aliquot withdrawal, analyte measurement, data processing, result evaluation, and decision-making, without requiring the assistance of an analyst. It also enables the restarting of the entire process to adjust it to the specific requirements of a new sample, if necessary. The positive feedback loop, which allows at least one connected operation to be regulated without human involvement, is the distinguishing characteristic of automated systems. In the automation of HPTLC analysis, different peristaltic pumps or other liquid management devices are used. Computer-controlled procedures have brought many benefits to HPTLC systems, among them a marked reduction in the amounts of samples and chemicals required. They have also made it possible to introduce the idea of unit laboratory activities. The automation of HPTLC has increased its importance in different analytical fields due to easy process, analyzing and handling. Further, other hyphenation techniques are adopted for better analysis and detection, which are discussed below [1].

5. Hyphenations

The hyphenated approach incorporates a chromatographic separation technique on-line with an efficient spectrometric or spectroscopic detector to analyze several chemical, biological and toxicological samples simultaneously, which has a wide variety of structural complexity. These hyphenation approaches have an essential role in the detection, identification and authentication of different mixtures of compounds in a minimal period. Because of their fast action, effectiveness and requirement of minimal amounts of material for analysis, these hyphenation techniques are preferred over the conventional analytical (single compound analysis at a time) methodology. Most of the time, such hyphenation may be done using analytical tools readily accessible on the market. The hyphenation of HPTLC is used for most (trace-level) analytical challenges, which assists in resolving various analytical issues and collecting more comprehensive analytical information from the samples [20, 21].

HPTLC is a significant alternative technology to high-performance liquid chromatography (HPLC) or gas chromatography (GC) due to the utilization of newly developed accessories like new chromatographic chambers, video scanners, densitometers, high-resolution sorbents with predetermined particle sizes or chemically modified surfaces, efficient elution techniques and it is potent to integrate with other analytical instruments like MS and the introduction of advanced computer programs for method optimization [22]. HPTLC-based hyphenation is favored over other hyphenated techniques because it presents fewer challenges than others. It has different advantages - being very sensitive to various sensitivity levels, reconfigurable instrumentation is more affordable [23], and readily reachable for the appropriate ideal solvent. After the process, the eluent is removed and does not affect the various detectors. The primary distinction is that the samples are often eluted as waste following HPLC separation. However, in HPTLC, the sample is still in the stationary phase (silica) and accessible for further experimental work and helps gather information with minimal sample wastage [23]. For a single separation, multiple spectrometers were successively employed like- FTIR or attenuated total reflection (ATR), UV, Vis, fluorescence detection (FLD), surface-enhanced Raman spectroscopy (SERS), mass spectrometry (MS) or high-resolution MS (HRMS). The essential existing hyphenations are:

5.1. LC-HPTLC

High-performance liquid chromatography is the best analytical tool for detecting and identifying different samples present in a mixture according to their retention time. However, the different samples can't be separated as they are eluted as a mixture. So, by adding simple HPTLC techniques, we can easily separate other mixture constituents already detected by the HPLC method according to their retardation factor (R_f). For this hyphenation, a simple interface is needed. The spray-jet applicator has been modified in various ways to provide a universal interface for connecting column liquid chromatography to thin-layer chromatography (LC-TLC). The mobile phase flow rate applied to the layer must be changed, so a splitter is employed in the transfer line to the spray-jet applicator to handle a high flow rate from wider-bore columns. Then nitrogen gas is mixed with the column eluent and the formed aerosol is applied to the chromatographic plate. Within a predetermined bandwidth, the head of the sprayer is moved horizontally along a single line [1]. This helps in separating, detecting, identifying and estimating compounds purity at a time.

5.2. HPTLC-MS

To collect analytical information on differentiated chemicals straight from the HPTLC plates, HPTLC is used in conjunction with MS. HPTLC-MS is a quick and sensitive method for isolating and identifying compounds from a complex sample with high purity outcomes and precise identification of compounds down to the nanogram and picogram concentrations [24]. In this technique, the concerned zones of the sample are placed into an MS for identification and the required data is obtained from MS within a minute. TLC is a method for substance preservation chromatography that keeps samples safe in a silica solid surface [25]. The mass spectrometric analysis adds another level of detail to the chromatographic procedure, increasing the specificity of detection and identification confidence [26]. Compared to LC-MS, HPTLC-MS has developed more slowly. The range of applications for which HPTLC is thought to be acceptable is constrained, thus, producing an automated system that can get direct mass spectra from the needed zone of the stationary phase directly is not an easy issue.

HPTLC-MS approaches may solve quality control problems in many medications, such as mixed herbal pharmaceuticals using their fingerprint spectra region. The target material from the HPTLC plate is directly eluted and transported on-line to the MS. On HPTLC plates, component combinations may be separated economically even with a high matrix load and if the target zone or desired sample is not visible, it can still be marked by extrapolating the nearby zone made visible by derivatization or by marking it under UV. The interface of HPTLC–MS is worked semiautomatically, where the estimated zone is positioned manually then the piston is released by pressing a button. After that, the lever is pulled to begin the sample extraction due to the solvent flow through the desired zone. Then the extracted solvent with the sample was introduced to the MS system, here different ionization techniques and detectors are used for the analysis of the sample, but before the introduction of the HPTLC-MS interface, data collection process had to be initiated by flow injection analysis, direct flow infusion, placebo injection, using data acquisition window. The cleaning process was then carried out. The HPTLC can also be hyphenated and incorporated into HPLC-MS. Here the hyphenation was done using a simple interface that contained the reverse phase C18 column with the appropriate guard column attached to the auto TLC- LC-MS interface, which was outfitted with a rounded elution head. A mass spectrometer was linked to the column. Both the hyphenations have recently been used effectively for various purposes, including the characterization and quantification of amino acids in peptides, the quick separation and characterization of unidentified impurities, the identification of botanicals, and the screening for bioactive natural substances [1]. Figure 3 describes instrumentation of HPTLC-MS.

5.2.1. Other Online Hyphenations Interfaces of HPTLC-MS

To overcome different challenges, several experiments were carried out and as a result some interface mechanisms were approved, which resolves the problems. The interface has been discussed below. The necessary plate is developed, followed by drying the plates using the flow of warm air and they are scanned with LC scanner (CAMAG). The system documented the estimated data. Primarily, chosen software was used to control all processes and instruments (WinCATS software 1.4.2 Planar Chromatography Manager (CAMAG)).

The plunger, a planar motor, a web camera, and an automated six-port valve are the four main parts of the automated interface, as shown in Figure 1. The plunger was attached to a linear actuator (z-axis), the Axial motor is mounted on the x and y axes of a 20 cm by 20 cm air plate that is suspended 10 mm above a lower structured surface (50 cm x 50 cm), Web camera was used to capture a plate picture, which was then uploaded to software to determine the plate's position (x, y-axis) and the automated six-port valve used to control the flow rate and pressure of the solvent. According to their need, different plungers are interchanged automatically with the help of plunger assembly. A series of solenoid valves with a fixed linear resolution was used at the bottom of the moving plate to position the planar motor in the x and y directions. To maintain the z-positioning, the ball screw spindle was used, and a motor monitored this. The ideal arrangement of all hardwires was entirely controlled by a computer, and the web camera photos were used to verify that the apparatus was placed correctly. Any band positioned on any type of plate size (up to 20 cm x 20 cm) in addition to various supports like glass or aluminum and thicknesses of layer (up to 200 or 100mm) may be evaluated using this interface. As soon as the plate was put on the air cushion plate, the program subsequently determined the precise location of the bands selected using the plate picture on the computer screen and the elution pattern was created by selecting the elution sequence.

The program also regulates the elution period, including the plunger cleaning process. The solvent flow rate was delivered using a pump, and different flow rates were set. Once the sample band to be analyzed was decided the plunger firmly positioned itself over the selected band and for the extraction process, the six-port valve automatically switched accordingly. Then the sample was eluted from silica and pumped through an inlet capillary using a solvent mixture. In the end, the outlet capillary and six-port valve were used to pump the extracted eluent into the various MS inlet (Z-spray ESI Quattro LCZ triple quadrupole mass spectrometer from Micromass) [27].

Then the sample was analyzed by using a different analyzer and the data was stored in the system. To avoid cross-over between two elution phases, the plunger was relocated to a docking station where it was mechanically cleaned (at least 4 s) using compressed air. After that, the blanks were eluted in between bands of samples and standards. Then the same process was repeated for another analytical process.

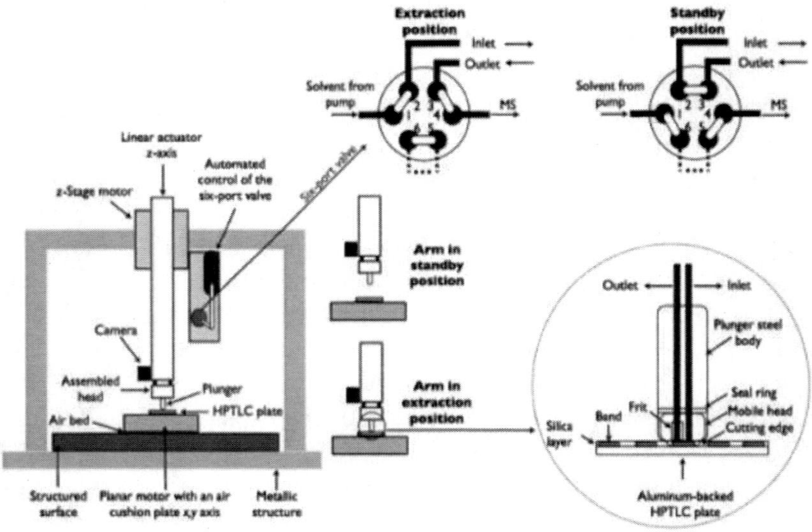

Figure 3. Instrumentation of HPTLC-MS.

5.3. HPTLC-NMR

NMR spectroscopy may also be used to characterize compounds that have been separated via HPTLC. The desired separated compound was extracted from the HPTLC plate, then the extracted material was mixed with D_2O_2 solution then the NMR process was carried out. The results demonstrate that analytes separated on an RP TLC plate can provide NMR spectra that can be interpreted without laborious isolation methods [28].

5.4. HPTLC-FTIR

Because HPTLC is reliable and easy to use, there is a demand for detection methods that enable sample identification and determination. IR spectroscopy has a high potential for identifying molecular structures and distinctive absorption bands. Nearly all chemical compounds produce excellent IR spectra, better for distinguishing between closely related molecules and identifying unknown substances. So infrared (IR) is coupled with HPTLC. Two categories of HPTLC and FTIR coupling are indirect and direct techniques. The indirect coupling includes either the desired chemical is extracted from the silica of the TLC plate and mixed with an IR material that doesn't absorb IR light like KBr, KCl or when the spectra are captured immediately from the plate, *in-situ* measurement of excised HPTLC spots. The more refined and straightforward technique measures the FTIR spectrum of the solute immediately (*in situ*, biological, pharmaceutical, environmental) on the plate [29], which is based on the online hyphenation of HPTLC-FTIR. The only hyphenation available online prior to this one was HPTLC-UV, which could not identify many UV inactive compounds. However, the HPTLC-FTIR spectra enables the identification of even non-UV absorbing materials on HPTLC plates. Because of these factors, the HPTLC-FTIR approach is more widely applicable [1]. The diffuse reflectance infrared Fourier transform (DRIFT) device, coupled to a Bruker IFS 48 FTIR spectrometer, is the fundamental component of the HPTLC-FTIR on-line hyphenated approach. It scans the plate with an infrared beam, shown in Figure 2. The plate is secured to an electronic x-y stage. The external interface mirror is attached to the spectrometer, which directs the infrared beam to the parabolic mirror by passing it through a single-sided mirror, so the reflected IR beam is referred to the plate at an angle of 90°. Most Fresnel reflection is not observed because the parabolic mirror returns the direct surface reflectance via the aperture back into the spectrometer (because it is eliminated in the 3,600–1,350 cm^{-1} region). At the parabolic mirror's edges, diffuse reflectance is collected and reflected the elliptical mirror through the one-sided mirror. Then the diffuse reflectance is collected and directed to the MCT (mercury, cadmium, telluride) detector. Software is used to obtain a chromatogram from an HPTLC plate. Peak purity may be quickly checked using this technique. The use of this technique in a variety of analytical domains, including forensic, culinary, environmental, and biological analysis, etc. [30]. Figure 4 describes instrumentation of HPTLC-FTIR.

5.5. HPTLC-Resonance Raman Spectrometry

It is necessary to have a homogenous surface for immediate and accurate analysis on the chromatographic plate following separation when using thin-layer chromatography (TLC) with surface-based spectrum techniques. Surface-enhanced Raman spectroscopy (SERS) is a compassionate and efficient analytical technology. Because of its enhanced sensitivity, this technique can identify different organic molecules because it can adsorb even if the organic samples are deposited on rough metallic surfaces [28]. It can be coupled with liquid chromatographic systems such as CE-TLC-SERS (31) or LC-TLC-SERS (32, 33) for a more convenient process. Background noise is not a concern because TLC plates typically are poor Raman scatterers. However, the Raman signals will be feeble on a bare TLC surface, because neither chemical nor electromagnetic SERS-enhancing methods work for amplification [31].

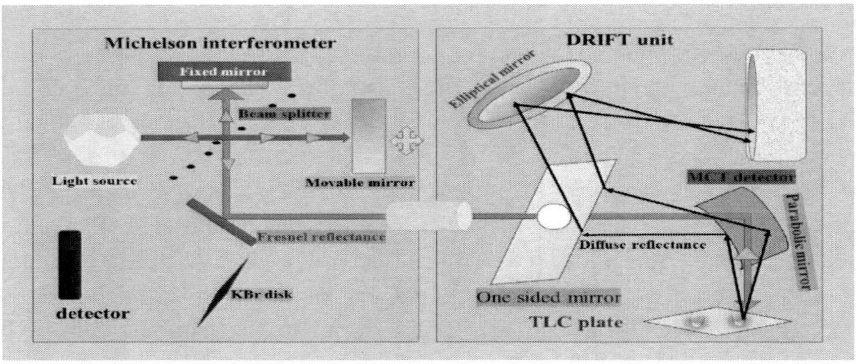

Figure 4. Instrumentation of HPTLC-FTIR.

This straightforward hyphenation method contains SERS, a specific Raman chromatographic thin layer and aluminum backplate Interface. Although the Raman TLC was created in the conventional method, several changes were made to assist in creating the interface. The processes involved developing the conventional TLC, then the mobile phase contents were evaporated at room temperature, then the stains were visualized, and silver sol (silver colloid) was applied to each stain or spot to be analyzed. The methyl Red was used as a marker for the analyte-modified aluminum backplate interfacial transfer. HPTLC-SERS's sensitivity allows it to identify substances in nanogram quantities [32]. The TLC-SERS hyphenation was used for a wide variety of solutes such as nitrogen-containing polyaromatic hydrocarbons (PAHs), acridine orange, cetylpyridinium chloride and amino triphenyl-methane dyes. Because the silver sols are reproducible, the requirement of the preparation of silver sols were significantly lower compared to other quantitative analytical processes. Therefore, TLC-SERS has been suggested to make it easier to identify solutes than to quantify them on the plate (31) precisely.

Conclusion

HPTLC is a versatile and efficient technique for qualitative as well as quantitative analysis of analytes with high sensitivity and accuracy. It has been widely adopted and utilized by several industries and researchers for analytical purposes. Its hyphenation with other techniques like Mass spectrometry, FTIR, NMR, etc. has broadened its adaptability to several folds due to improved detection properties.

References

[1] Srivastava, M. (2011). An overview of HPTLC: A modern analytical technique with excellent potential for automation, optimization, hyphenation, and multidimensional applications, *High-performance thin-layer chromatography (HPTLC)*, 3-24.

[2] Attimarad, M., Ahmed, K. M., Aldhubaib, B. E., and Harsha, S. (2011). High-performance thin layer chromatography: A powerful analytical technique in pharmaceutical drug discovery, *Pharmaceutical methods*, 2, 71-75.

[3] Zlatkis, A., and Kaiser, R. E. (2011). *HPTLC-high performance thin-layer chromatography*, Elsevier.

[4] Gökbulut, A. (2021). High Performance Thin Layer Chromatography (HPTLC) for the investigation of medicinal plants, *Current Analytical Chemistry*, 17, 1252-1259.

[5] Borgerding, M. F., Hinze, W. L., Stafford, L. D., Fulp, G. W., and Hamlin, W. C. (1989). Investigations of stationary phase modification by the mobile phase surfactant in micellar liquid chromatography, *Analytical Chemistry*, 61, 1353-1358.

[6] Bolliger, H. R., Brenner, M., Gänshirt, H., Mangold, H. K., Seiler, H., Stahl, E., and Waldi, D. (1965). *Thin-layer chromatography: a laboratory handbook*, Springer.

[7] Sonia, K., and Lakshmi, K. (2017). HPTLC method development and validation: An overview, *Journal of Pharmaceutical Sciences and Research*, 9, 652.

[8] Cañigueral i Folcarà, S., Frommenwiler, D., Reich, E., and Vila Casanovas, R. (2018). High performance thin-layer chromatography (HPTLC) in the quality control of herbal products, *Recent Advances in Pharmaceutical Sciences VIII*, 2018,

Research Signpost. Editors: Diego Muñoz-Torrero, Yolanda Cajal and Joan Maria Llobet. ISBN: 978-81-308-0579-5. Chapter 7, p. 119-136.
[9] Argekar, A. (2006). *Planar Chromatography in Pharmaceutical Analysis, Encyclopedia of Analytical Chemistry: Applications, Theory and Instrumentation.*
[10] Rashmin, P., Mrunali, P., and BHARAT, P. (2012). HPTLC method development and validation: strategy to minimize methodological failures, *Journal of food and drug analysis*, 20.
[11] Patel, R. B., Patel, M. R., and Batel, B. G. (2011). Experimental aspects and implementation of HPTLC, *High-performance thin-layer chromatography (HPTLC)*, Springer, pp. 41-54.
[12] Sherma, J. (2010). Review of HPTLC in drug analysis: 1996-2009, *Journal of AOAC international*, 93, 754-764.
[13] Ambati, P., Bala Krishna, P., Srinivasa Rao, Y., Varaprasada Rao, K., and Deepthi, R. (2021). *Instrumentation and Future Prospects of HPTLC-A.*
[14] Wagner, H., and Bladt, S. (1996). *Plant drug analysis: a thin layer chromatography atlas*, Springer Science & Business Media.
[15] Renger, B. (1993). Quantitative planar chromatography as a tool in pharmaceutical analysis, *Journal of AOAC International*, 76, 7-13.
[16] Kulkarni, R., Pandhare, R., Deshmukh, V., Mohite, P., and Pawar, A. (2021). High-performance thin layer chromatography: A powerful analytical technique in pharmaceutical drug discovery, *Journal of Pharmaceutical and Biological Sciences*, 9, 7-14.
[17] Hussain, S. Z., Maqbool, K., and Naseer, B. (2019). High performance thin layer chromatography: Principle, working and applications, *Int J Res Pharm Pharm Sci*, 4, 83-88.
[18] Saibaba, S., and Pandiyan, P. S. *Research in Pharmacy and Health Sciences.*
[19] Jain, A., Parashar, A., Nema, R., and Narsinghani, T. (2014). High performance thin layer chromatography (HPTLC): A modern analytical tool for chemical analysis, *Current Research in Pharmaceutical Sciences*, 8-14.
[20] Wilson, I., and Brinkman, U. T. (2003). Hyphenation and hypernation: the practice and prospects of multiple hyphenation, *Journal of Chromatography A*, 1000, 325-356.
[21] Srivastava, M. (2010). *High-performance thin-layer chromatography (HPTLC)*, Springer Science & Business Media.
[22] Sharma, B., Islam, A., and Sharma, A. (2023). *HPTLC-MS: an advance approach in herbal drugs using fingerprint spectra and mass spectroscopy.*
[23] Brosseau, C. L., Gambardella, A., Casadio, F., Grzywacz, C. M., Wouters, J., and Van Duyne, R. P. (2009). Ad-hoc surface-enhanced Raman spectroscopy methodologies for the detection of artist dyestuffs: thin layer chromatography-surface enhanced Raman spectroscopy and *in situ* on the fiber analysis, *Analytical Chemistry*, 81, 3056-3062.
[24] Ramu, B., and Chittela, K. B. (2018). High Performance Thin Layer Chromatography and Its Role Pharmaceutical Industry, *Open Sci. J. Biosci. Bioeng*, 5, 29-34.

[25] Juszczak, A. M., Zovko-Končić, M., and Tomczyk, M. (2019). Recent Trends in the Application of Chromatographic Techniques in the Analysis of Luteolin and Its Derivatives, *Biomolecules*, 9, 731.

[26] Verma, K. L., Kumar, M., and Singh, A. P. (2018). HPTLC-MS as a Neoteric Hyphenated Technique for Separation and Forensic Identification of Drugs, *Journal of Analytical Sciences, Methods and Instrumentation*, 8, 1-15.

[27] Luftmann, H., Aranda, M., and Morlock, G. E. (2007). Automated interface for hyphenation of planar chromatography with mass spectrometry, Rapid Communications in Mass Spectrometry: *An International Journal Devoted to the Rapid Dissemination of Up-to-the-Minute Research in Mass Spectrometry*, 21, 3772-3776.

[28] Tuzimski, T. (2011). *Basic principles of planar chromatography and its potential for hyphenated techniques, High-Performance thin-layer chromatography (HPTLC),* Springer, pp. 247-310.

[29] Cserháti, T., and Forgács, E. (2020). Hyphenated Techniques in Thin-Layer Chromatography, *Journal of AOAC International*, 81, 329-332.

[30] Cimpoiu, C. (2005). Qualitative and Quantitative Analysis by Hyphenated (HP) TLC-FTIR Technique, *Journal of liquid chromatography & related technologies*, 28, 1203-1213.

[31] Oriňák, A., Talian, I., Efremov, E. V., Ariese, F., and Oriáaková, R. (2008). Diterpenoic acids analysis using a coupled TLC-surface-enhanced Raman spectroscopy system, *Chromatographia*, 67, 315-319.

[32] Workman, J. (1993). A review of process near infrared spectroscopy: 1980–1994, *Journal of Near Infrared Spectroscopy*, 1, 221-245.

Chapter 7

Two-Dimensional Liquid Chromatography: Instrumentation and Applications

Ramesh Kohar[1,*]
and Amandeep Singh[2,†]
[1]Department of Pharmaceutical Analysis, ISF College of Pharmacy,
Ghal Kalan, Moga, Punjab, India
[2]Department of Pharmaceutics, ISF College of Pharmacy,
Ghal Kalan, Moga, Punjab, India

Abstract

Two-dimensional liquid chromatography has been widely used to separate and analyze various compounds by different analytical researchers. As it offers high resolution of very complex samples, it has been used diversely in analytical research. This chapter included detailed 2D-LC, its working principles, advantages and disadvantages, and instrumentation. Beyond this, we also comprise numerous applications in different fields such as pharmaceutical, biopharmaceutics, food, natural medicines, environmental, metabolites and many other analyses that many analytical researchers study. 2D-LC offers sensitive, selective, reproducible, rapid separation and analysis with high resolution, orthogonality and high peak capacity.

Keywords: 2D-LC, analytical, solvent, pump, column, detector

* Corresponding Author's Email: koharramesh69@gmail.com.
† Corresponding Author's Email: ad4singh@gmail.com.

In: Modern Approaches in Fluid Chromatography
Editors: Rohit Bhatia and Bhupinder Kumar
ISBN: 979-8-88697-856-8

1. Introduction

Two-dimensional liquid chromatography (2D-LC) is a technique that uses two independent column systems to separate a complex mixture of samples utilizing two different liquid phase separation techniques with high resolution and high peak capacity (one second per peak). In this technique, the effluent from the first dimension column (1D column) is allowed to pass onto the second dimension column (2D column) either by portions called heartcutting 2D chromatography or by the entirety in a sequential manner with small aliquots called comprehensive 2D chromatography [1-3]. It is a hyphenated technique in which two liquid chromatography systems are coupled by the interface (valve and storage loop) [1, 4, 5]. The chromatogram obtained is the retention time versus peak intensity (detector response) plot of the first dimension on the x-axis and the second on the y-axis [6].

The benefits of two-dimensional liquid chromatography are that the separation of components from a very complex mixture can be performed quickly and rapidly with high resolution that can not be separated by one-dimensional liquid chromatography (1D-LC) and offers high efficiency and selectivity among homogenous pairs of components (like enantiomers, isomers) in the complex mixture of a sample with high peak capacity [7-9].

The disadvantage of two-dimensional liquid chromatography is that the method development is challenging since the retention mechanism of two different column systems is comparatively different [10, 11]. Another limitation may be the undersampling (i.e., either sampling a small volume aliquot than the required peak volume or more volume aliquot than the peak needed) [11, 12]. After each pass of an aliquot of effluent, column re-equilibration is required since only one fraction from one dimension column is allowed to pass through the second dimension column at a time interval for complete and pure separation [13, 14]. Another problem is solvent incompatibility in the case of two different retention mechanisms and columns [15].

The difference between 1D-LC and 2D-LC is that 1D-LC comprises only one column for separating a non-complex mixture of samples, but 2D-LC comprises two different columns of liquid chromatography for the separation and resolution of a complex mixture of the sample. The 1D-LC takes more time to break the sample with low peak capacity, but 2D-LC rapidly separates complex mixtures with high peak capacity. 1D-LC has low efficiency than 2D-LC [1, 10, 16]. The sample containing more than 50 analytes may cause overlapping peaks even with the most efficient 1D-LC, but this problem is resolved using 2D-LC [8, 17]. Figure 1 describes the typical instrumentation of 2D-LC.

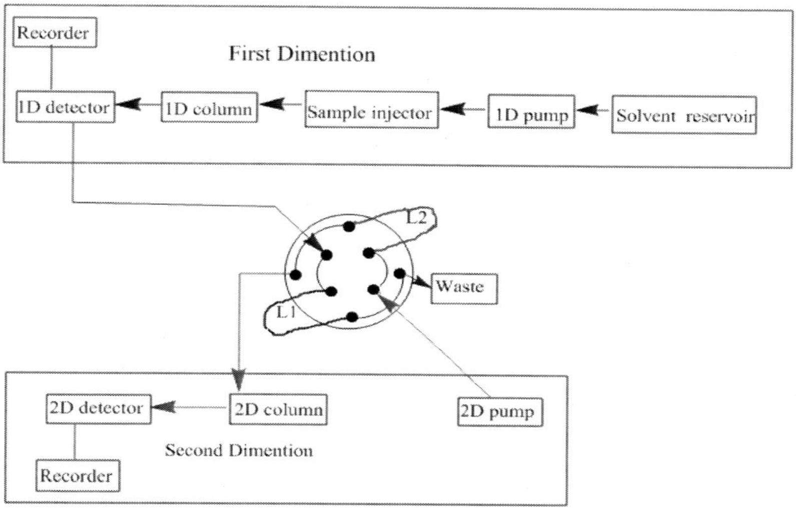

Figure 1. Schematic diagram of 2D-LC instrumentation.

2. Instrumentation

There are two types of two-dimensional liquid chromatography [3, 5, 6, 18]:

1. Heartcutting 2D liquid chromatography allows only a small portion of 1D effluent to pass through the 2D column system for accurate and pure separation of components.
2. Comprehensive 2D liquid chromatography allows all fractions of 1D effluents to pass through the 2D column in many fine aliquots to separate the complex mixture components rapidly.

Followings are the components of 2D-LC [19, 20];

- Solvent reservoir
- Pump system
- Sample injector
- Column

- Detector
- Recorder and amplifier
- Interface (8 or 10 port valve system and storage loop system) [21, 22]

2.1. Solvent and Reservoir [15, 23-27]

The reservoir may be stainless steel, plastic, or glass vessel of capacity ranging from 100 ml to several litres, which is used to store the solvent (mobile phase). Glass vessel is most commonly used as a reservoir.

The solvents used should be suitable for columns of both dimensions. If the first column is a normal phase column and the second is a reversed-phase column, then the solvent used should be polar to facilitate fast separation in the second dimension column. Thus solvents to be used are water, acetone, acetonitrile, ethanol, methanol, isopropanol, tetrahydrofuran, buffer solvents etc.

2.2. Pump [5, 9, 20, 24, 28-31]

The pump transfers the solvent (mobile phase) into the column with the required pressure and controlled flow rate. The flow rate for the first dimension is relatively slower than the second dimension to allow sufficient separation of components in the second dimension. The pump in the first dimension is installed just before the column, whereas the second dimension, established before the second storage loop, injects content into the second-dimension column. The following three types of pumps are most commonly used in liquid chromatography (2D-LC).

2.2.1. Reciprocating Pump

This type of constant flow pump uses an electric motor to drive the piston for transferring the mobile phase. The piston move in alternate motion in the direction of the frow and back that help to push and pull of mobile phase to the column and from the reservoir respectively. Due to the pulsating flow rate of this pump, the damper is used on the pump outlet to reduce pulsation. Even the dual-piston pump can be used to reduce pulsation. The reciprocating pump can withstand pressure up to 10,000 psi and deliver 5-250 μl/min, but the internal volume capacity is up to 50 μl.

The reciprocating pump can be divided into:

- Single piston pump
- Parallel double-piston pump
- Serial double-piston pump

A single-piston pump consists of a single piston that drives the mobile phase in one direction using a check valve system (Figure 2).

In the parallel double-piston pump, if one piston drives the mobile phase into the column, another piston refills the chamber from the reservoir (Figure 3).

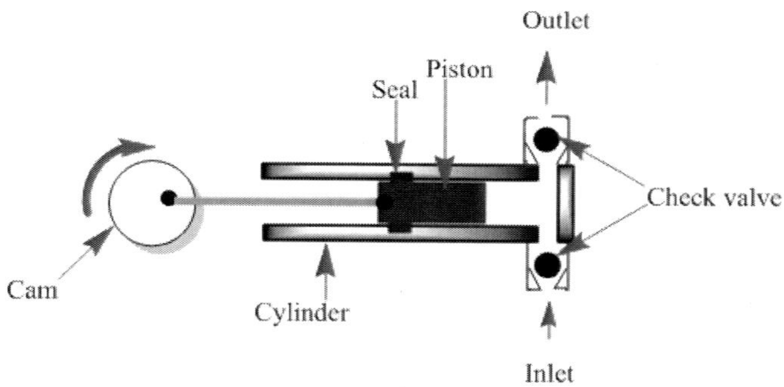

Figure 2. Single piston reciprocating pump.

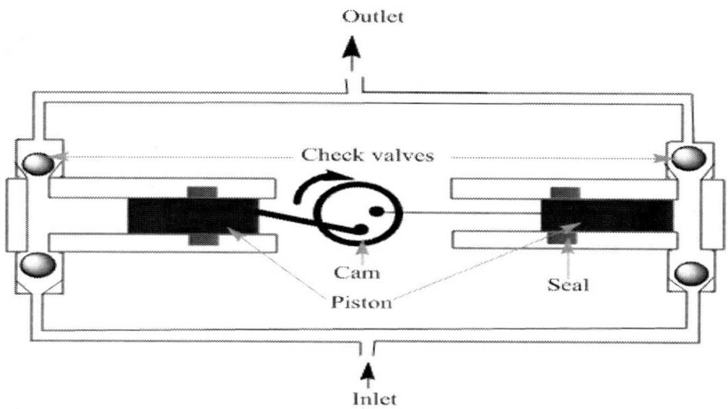

Figure 3. Parallel double piston reciprocating pump.

The mechanism of the serial double-piston pump is the same as the parallel one. If one piston drives the mobile phase into the column, the other piston refills the chamber from the reservoir. As the piston reaches the end, the mechanism reverses its direction to fill the mobile phase from another piston chamber (Figure 4).

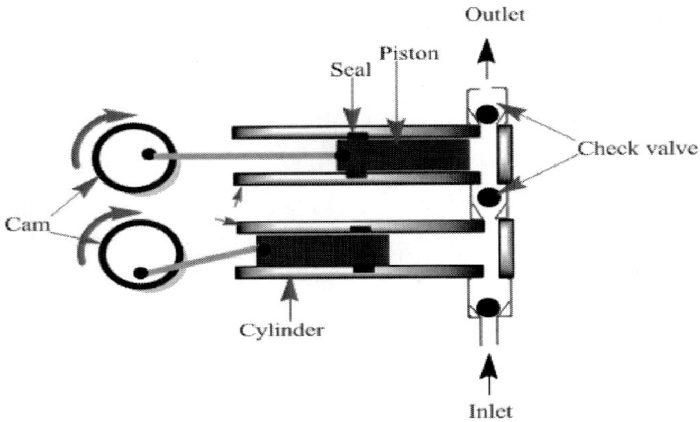

Figure 4. Serial double piston reciprocating pump.

2.2.2. Syringe Type Pump or Screw Driven Pump

This is a constant flow pump in which the electric stepper motor helps to move a piston through a screw which transfers the mobile phase to the column from the reservoir. The advantage of this pump is that the flow rate is pulseless, but the internal volume is limited, which should be refilled if the mobile phase reservoir is empty. Two syringe pumps are connected to overcome this problem. Thus, if the mobile phase reaches the end of the first syringe pump, the activation for delivery by the second pump is done and the refilling of the first syringe pump is started. The time for stabilizing the flow rate by syringe pump is long, which depends on the flow rate, chamber volume of the pump, the compressibility of fluid and column permeability. It can withstand a pressure of more than 14,500 psi and the volume capacity is up to 10 ml. The flow rate ranges from 20-2000 nl/min.

2.2.3. Pneumatic Pump

A pneumatic pump is a constant-pressure pump that drives the mobile phase using gas. There are two types; one is a pneumatic pump without amplification having a reservoir for the mobile phase constructed with an outlet for the mobile phase and a gas inlet. Thus, the pressure experienced by the mobile phase equals the gas pressure used to pump the mobile phase onto the column (Figure 5). The simple construction, low cost, and pulseless flow rate are the advantages of this type of pump, but the limitations are limited working pressure below 2900 psi (i.e., below 200 bar) and interruption to refill the reservoir by the mobile phase.

Another is a pneumatic pump with amplification in which a piston gas exerts pressure to drive the mobile phase. The contact area of the piston to the mobile phase is lower but with the gas is higher, allowing pressure amplification. Hence, considering that 10 times larger the diameter of the piston contacting with gas is than that of the piston reaching the mobile phase, the pressure of the mobile phase can be 100 times the pressure of the gas used. The advantage of this pump is uninterrupted pumping with more significant pressure using lower pressure gases, but the limitations are a generation of pulsed flow and the inability of gradients generation.

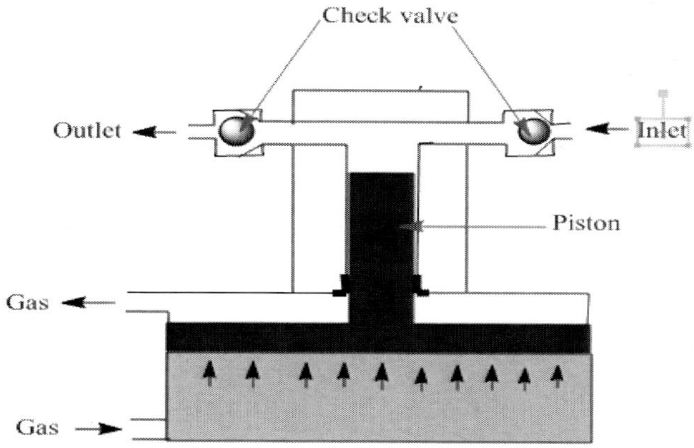

Figure 5. Pneumatic pump.

2.3. Sample Injector [5, 20, 22, 24, 32, 33]

These are used to inject the sample solution onto the column for analysis. The sample injector is installed between the pump and column in the first dimension only, not in the second dimension since the modulating interface acts as the injector in the second dimension. Three types of sample injection systems are available that are as follows:

2.3.1. Septum Injector

In this type of injector, the high-pressure syringe is used to introduce the sample through the self-sealing septum, but the limitation is the leaching effect in contact with the septum, which leads to the formation of ghost or pseudo peaks. Thus, this type of pump is commonly not used.

2.3.2. Stop-Flow Injector

This a septumless injector. Hence most of the problems associated with septum-type injectors are eliminated. In this type of injector, the mobile phase flow is stopped for some time and the top of the column is opened if the ambient pressure reaches and then onto the top of the column sample is injected. The volume capacity is up to 5 µl. These are recently not used because they may cause pressure drop and generate pulse due to the stooping of mobile phase flow during the injection period.

2.3.3. Loop Injector or Rheodyne Injector

Most commonly used injector for precise volume injection. This is sometimes called a microvolume sampling valve type injector, which consists of a sample loop for storage of sample to be injected onto the column through the mobile phase without interruption of the mobile phase flow at high pressure (Figure 6). This is also adaptable for automatic injection. The sample volume for the automated system ranges from 2 µl to 100 µl. But for manual injection, the volume capacity of the rheodyne injector is available in 20 µl, 40 µl, 50 µl, 100 µl etc.

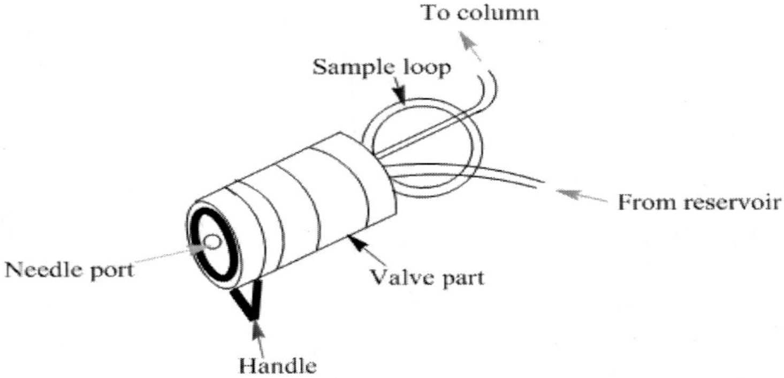

Figure 6. Rheodyne injector.

The sample loop operates in two modes, i.e., (a) sampling mode, i.e., load position in which the sample of required volume is loaded into the external sample loop and (b) injection mode i.e., inject position in which the loaded volume is injected onto the column through injection into the mobile phase.

2.4. Column

The most commonly used column dimensions are of length range from 50 – 300 mm and an internal diameter range from 1-4.6 mm with a particle size of stationary phase range from 1-5 μm. The column comprises stainless steel or glass but is most commonly made of stainless steel for its high strength and water-resistant properties. The internal diameter of the first dimension column is relatively narrower than the second-dimension column and the length of the first column is more excellent than the second-dimension column to facilitate rapid separation in the second dimension but slower separation in the first dimension [1, 34, 35]. Table 1 summarizes some common 2D columns. Table 2 describes stationary phases for 2D-liquid chromatography.

Table 1. Few combinations of two-dimensional columns [4, 8, 10, 13, 36-51]

First dimension	Second dimension
IEX	RPLC
SEC	RPLC
AgLC	RPLC
NPLC	RPLC
RPLC	RPLC
HILIC	RPLC
HILIC	HILIC
RPLC	Chiral
SEC	NPLC
SEC	IEX
Affinity	SEC
SEC	CZE

SEX: Ion exchange chromatography, RPLC: Reversed-phase liquid chromatography, SEC: Size exclusion chromatography, NPLC: Normal phase liquid chromatography, HILIC: Hydrophilic interaction liquid chromatography, AgLC: Argentation liquid chromatography, CZE: Capillary zone electrophoresis.

Table 2. Commonly used stationary phase for different columns [10]

Columns	Stationary phases
RPLC	C8, C12, most commonly C 18 etc.
IEX	Sulfonic acid, carboxylic acid, triethyl amine, quaternary amine
SEC	Crosslinked poly (styrene- divinyl benzene), methacrylate porous beads, porous silica etc.
NPLC	Silica, alumina, amino-propyl etc.
AgLC	Sulfonic acid or bare silica loaded with silver ions
HILIC	Zwitterionic: sulfobetaine, phosphocholine; Basic: amino propyl and Neutral: diol, amide etc.
Chiral	Polysaccharide derivatives (chiral carbamate/benzoate polymers of cellulose and amylose)
Affinity	Cephadex, polyacrylamide etc., with chemically bonded antigens or antibodies

2.4.1. Mixer and Dilution Pump (Optional)

A mixer is required to mix the dilution solvent properly and the eluent from the first dimension if there is needed to dilute the eluent from the first dimension to facilitate retention of the analyte eluted from the first dimension column in the trapping column to be injected into the second dimension column (Figure 7). This is optional in the instrumentation. It should be installed between the first dimension detector and the modulation interface. If dilution is needed, then one extra dilution pump must be installed by connecting the dilution solvent reservoir and the mixer to pump the solvent for dilution into the mixer.

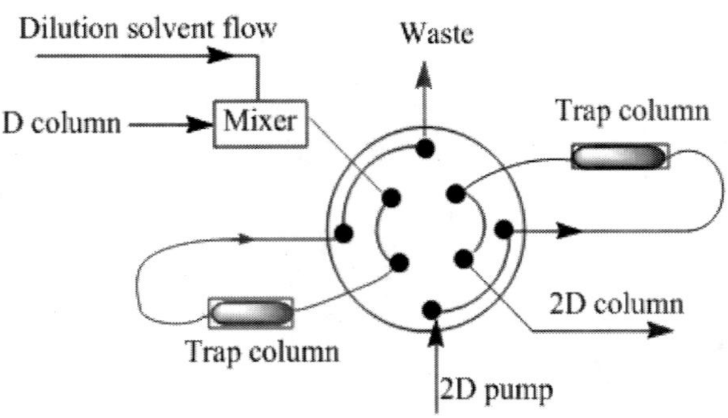

Figure 7. The modulation interface has a trapping column and is connected to the mixer and dilution pump in the instrumentation.

The pump must be the same as described in the section [8]. These are installed only when the available sample loop is replaced by the trap (small volume trapping column) in the modulation interface and usually the guard column consisting stationary phase similar to the second dimension column. This is also called stationary phase-assisted modulation in the interface [52].

2.4.2. Coupling Interface

The two dimension columns are coupled using a transfer interface (modulation interface), also called a modulator. The modulator is an eight or 10-port and 2-position switching valve connected with two similar storage loops by which, alternatively, one loop samples the first dimension eluent and the other loop's content is injected into the second-dimension column (Figure 8). The storage loop has a volume capacity of 10-50 μl range. The filling and injection cycle between two loops is controlled by regular valve switches [8, 21, 22, 53, 54].

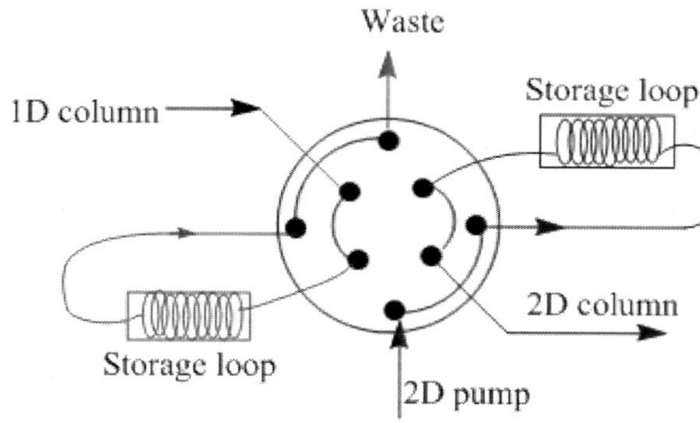

Figure 8. Port, two-position valve.

2.5. Detector

The detector is used to detect the separated analytes in the eluent. The detector is installed after the column in both the dimensions, i.e., in the first and second dimensions represented as ^{1}D detector and ^{2}D detector in the instrumentation figure. The detector should have a fast scanning capacity for the second dimension since it is operated at a very high speed. But the first dimensions operated at a relatively slow speed; hence, a slow-scanning detector may be used here. Following are the generally used detectors [6]:

- UV-visible detector and Photodiode array detector (PDA)
- Mass spectrometry (MS) detector

- Evaporative light scattering detector (ELSD)
- Fluorescence detector

2.5.1. UV-Visible Detector and Diode Array Detector (DAD)

The UV-visible detector is based on the absorption of UV-visible light by the eluent emerging from the columns and travelling through a photocell placed in the radiation beam. The absorption wavelength ranges from 200-400 nm for UV light and 400-800 nm for visible light. It can detect the samples like alkenes, aromatic compounds and compounds with multiple bonds that absorb UV-visible light specifically. In this case, the solvent must not absorb UV-visible radiation [24]. The flow cell has a volume capacity range from 1-10 μl with a 1 mm diameter and 2-10 mm path length. The parameters depend on the detector's instrumentation design [55, 56] (Figure 9).

PDA detector is a type of UV-visible detector in which polychromatic light is allowed to pass through the flow cell. A grating is used to diffract all outcoming light of radiation from the flow cell and then reaches an array of photodiodes where every photodiode gains a unique narrow band of wavelength (Figure 10).

The PDA can scan the array of photodiodes many times within a second. Thus, its application is increasing nowadays. It has high sensitivity and selectivity with multi-wavelength scanning properties [24, 55].

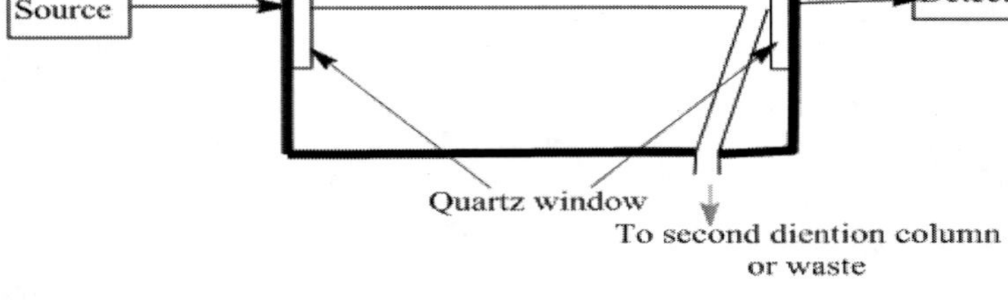

Figure 9. Schematic diagram of UV-visible detector.

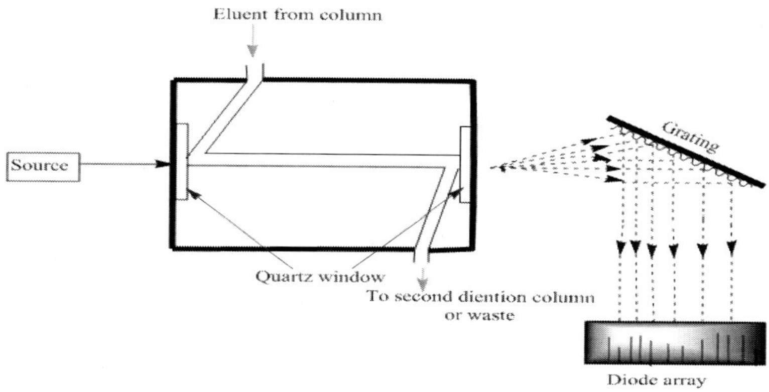

Figure 10. Schematic diagram of PDA detector.

2.5.2. Mass Spectrometry (MS) Detectors

MS detectors are used primarily when the 2D-LC is coupled with mass spectrometry, especially for the identification, mass determination, and structural elucidation of the eluent from the column analyte. It has high sensitivity, accuracy and resolution. The eluent of the 2D-LC is first ionized by an ionization source and then analyzed by an analyzer and detected by a detector like an electron multiplier etc. [6, 57].

2.5.3. Evaporative Light Scattering Detector (ELSD)

The principle of the evaporative light scattering detector (ELSD) is the evaporation of the mobile phase and then the measurement of light scattered by the dried particles of the sample. The eluent is nebulized with a stream of nitrogen or air carrier gas into a heated drift tube. Light scattered by a particle is detected by mounting a photocell at an angle to the incident beam of radiation. The sensitivity of the detector is decreased with volatile analytes and chromophoric analytes. The mobile phase must be volatile in nature [55, 58].

2.5.4. Fluorescence Detector

A fluorescence detector measures the light emitted by the analyte after excitation at a higher energy state (wavelength) by absorbing UV light. It is susceptible to fluorescent compounds and compounds that fluoresce by derivatization [59]. The eluent can pass through a flow cell irradiated with xenon or deuterium light source passing through the first filter (Figure 11).

The analyte in the eluent absorbs light and is then excited to a higher energy state, returning to a lower energy state and emitting radiation of higher energy wavelength, which is measured at the right angle to the incident beam of light after passing through the second filter [24].

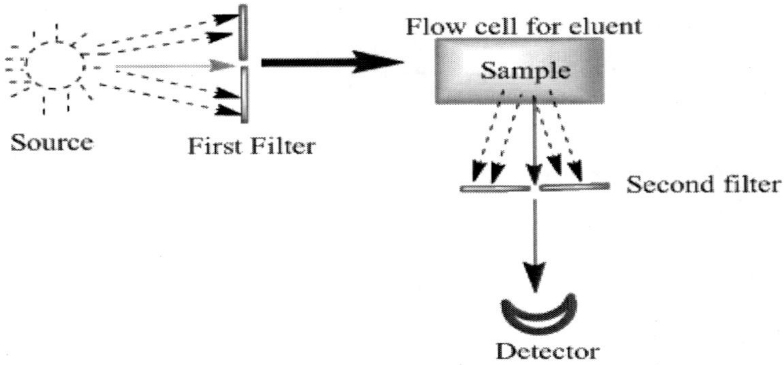

Figure 11. Schematic diagram of fluorescence detector.

2.6. Recorder

The recorder helps to record the data obtained from the detector response in the form of a chromatogram.

3. Applications

3.1. Food Analysis

3.1.1. Lipidic Food

Silver-ion high-performance liquid chromatography (HPLC) is a widely used advanced and versatile liquid chromatography for separating lipidic food by coupling with reversed-phase liquid chromatography (RPLC). Lipidic food is essential for our body to fulfil the requirement, maintain body structure, etc. L. Mondello et al., 2005 [25] used silver-ion reversed-phase comprehensive two-dimensional liquid chromatography combined with mass spectrometric detection in food analysis and analyzed triacylglycerol (TAG) in rice oil. This demonstrated that it is the most effective method for determining the TAG profile in lipidic food matrices. It can be a versatile tool for analyzing the different types of compounds with a wide range of polarity.

3.1.2. Carotenoids

The comprehensive normal phase liquid chromatography (NPLC)- reversed-phase liquid chromatography (RPLC) two-dimensional liquid chromatography is an advanced and widely used versatile mode of high-performance liquid chromatography for the analysis of various compounds in different food materials. Carotenoids are natural pigments used as a coloring agent in the food industry and these possess antioxidant properties [60-62], used for the prevention of cardiovascular disease [63, 64], cancer [65, 66] and muscular degeneration [67]. Paola Dugo et al., 2008 [68] studied the application of comprehensive two-dimensional liquid chromatography to elucidate the native carotenoid composition in red-orange oil. They analyzed free carotenoids and carotenoid esters from orange essential oil using the NPLC-RPLC-photodiode array detector (PDA)/mass spectrometric detector (MS) method. The identified free carotenoid and carotenoid monoesters were phytofluene, β-cryptoxanthin laureate, β-cryptoxanthin myristate, β-cryptoxanthin palmitate, lutein laureate. Lutein myristate, lutein palmitate etc. This type of liquid chromatography and methodology can be used to identify the intact carotenoids of other natural matrices.

3.1.3. Polyphenolic Foods

Hydrophilic interaction liquid chromatography (HILIC) is a type of modern and modified liquid chromatography widely used for separating compounds with a wide range of polarities in conjunction with reversed-phase liquid chromatography (RPLC). Polyphenols are the most crucial component that reduces the risk of atherosclerosis, cancer, and neurodegenerative disease [69], and viral infection [69]. Eduardo Somnella et al., 2017 [70] used online comprehensive hydrophilic interaction chromatography-reversed phase ultra-high-pressure liquid chromatography to analyze polyphenolic food samples. This study used an extract of the Annurca apple variety and found hydroxycinnamic acids, dihydrochalcones, anthocyanins, flavonols, flavanones, procyanidins etc. He demonstrated that this is the most efficient tool for separating and analyzing polyphonolic compounds in different types of foodstuff.

3.1.4. Carbohydrates

Carbohydrates are the most crucial components in food as a complex mixture of isomeric molecules with different degrees of polymerization, which differ only in their hydroxyl group configuration and glycosidic linkage position. Carbohydrates provide energy and dietary fibre to our bodies [71]. Comprehensive two-dimensional liquid chromatography is one of the most widely used modern liquid chromatography for rapid separation with high resolution of various compounds. Hydrophilic interaction liquid chromatography (HILIC) in conjunction with reversed-phase liquid chromatography (RPLC) is a type of two 2D-LC widely used for the analysis of compounds with different polarities like disaccharides and trisaccharides from probiotic oligosaccharides. Andrea Martin Ortiz et al., 2019 [72] studied the separation of di- and trisaccharide mixtures from prebiotic oligosaccharides by comprehensive two-dimensional liquid chromatography and separated the first-dimension coeluted di- and trisaccharides in second dimensions in the case of glucosyl-disaccharides; kojibiose was separated from isomaltase and cellobiose from laminaribiose in the second dimension, but the separation of maltose from cellobiose was done in the first dimension. He demonstrated that this new method is one of the most effective and efficient methods for the separation and analysis of di- and trisaccharides from different food materials.

3.1.5. Beer for Mycotoxins

Multiple heart-cutting two-dimensional liquid chromatography is a modern and versatile type for separating and analyzing compounds with vast polarity. Mycotoxins are toxic metabolites having low molecular weight produced by fungi like Aspergillus and Penicillium genera [73]. Luca Campone et al., 2020 [74] used multiple heart-cutting two-dimensional liquid chromatography-tandem mass spectrometry methods to determine mycotoxin in beer. They separated and identified aflatoxin G1, G2, B1, B2 and fumonisins B1, B2 by coupling C8 in the first dimension and kinetic biphenyl in the second dimension with a mass spectrometry detection system. He demonstrated that this method could be a rapid and automated multiple heart-cutting two-dimensional liquid chromatographic method for the determination of mycotoxins in beer.

3.1.6. Black Chokeberries Pomace

Hydrophilic interaction liquid chromatography (HILIC) is a type of modern and modified liquid chromatography widely used for separating compounds with a wide range of polarities in conjunction with reversed-phase liquid chromatography (RPLC). Black chokeberry contains chemical constituents like anthocyanin, proanthocyanidins, and polyphenols with potent antioxidant activity [75, 76]. Brazdausaks et al., 2016 [44] used comprehensive two-dimensional liquid chromatography with mass spectrometric detection-based chemical characterization of bioactive from black chokeberries pomace and analyzed and identified anthocyanins, proanthocyanidins, flavonoids and phenolic acids. He demonstrated this method as a new method for identifying constituents in black chokeberry pomace.

3.2. Applications in Biopharmaceutics

3.2.1. Free Drug Assay

Size exclusion chromatography is one of the advanced and versatile liquid chromatography for separating compounds with a wide range of molecular weight and size and with different polarity when used in conjunction with reversed-phase liquid chromatography. Antibody-drug conjugates are modern therapeutic agents that combine the selectivity of monoclonal antibodies and more efficacious minimal molecule drugs to eliminate tumor cells while reducing the therapeutic agent's general toxicity and side effects [77]. Yi Li et al., 2016 [37] use size exclusion chromatography-reversed phase chromatography as two-dimensional liquid chromatography for the free drug assay for antibody-drug conjugates and found that the antibody-drug conjugate recovery is less (<60%) as compared to linker drug (>90%) which demonstrate that the free drug is less. He demonstrated that this method could be one new approach for free drug assay for antibody-drug conjugate with automation and high throughput.

3.2.2. Analysis of Key Linker Drug Intermediate Used in Antibody Drug Conjugates

High-resolution two-dimensional liquid chromatography is an advanced form of high-resolution high-performance liquid chromatography in which both first and second-dimension columns are the same, i.e., reversed-phase liquid chromatography for the separation of a wide range of hydrophobic compounds. Antibody-drug conjugates are modern and advanced targeted therapeutics that can be used with radiotherapy and/or chemotherapy treatment to eliminate tumor cells. Antibody-drug conjugate is made with monoclonal antibodies and the linker drug component. Venkatramani et al.,

2017 [78] used high-resolution two-dimensional liquid chromatography to analyze linker drugs used in antibody-drug conjugate and found the purity of linker drug intermediate of 3 different lots in the range of 88.25 to 94.47%. He also demonstrated that this method could analyze structurally similar, coeluting impurities from linker drug intermediates used in antibody-drug conjugates in detail, which is not possible with one-dimensional liquid chromatography.

3.2.3. Analysis of Bovine Insulin

Two-dimensional liquid chromatography is a versatile type for separating very complex components. There is more growth in the application of 2D liquid chromatography for analyzing active pharmaceutical ingredients, excipients, drug products and biopharmaceutical materials [79-82]. Dwight R Stoll et al., 2017 [13] used two-dimensional liquid chromatography by coupling reversed-phase liquid chromatography in both first and second-dimension columns to analyze bovine insulin and found degradants in heat-treated bovine insulin characterized by different small peaks in the chromatogram.

3.2.4. Analysis of Oligonucleotide Impurities

Size exclusion chromatography is one of the advanced and versatile liquid chromatography for separating compounds with a wide range of molecular weight and size as well as with different ionization when used in conjunction with ion pair-reversed phase liquid chromatography. Reversed-phase liquid chromatography coupling with ion pair chromatography is a new two-dimensional liquid chromatography widely used for analyzing compounds with a wide range of polarity and ionization. Oligonucleotides have wide potential for treating various diseases and thus comprise a therapeutic agent platform. Stilianos G. et al., 2018 [41] used two-dimensional liquid chromatography-mass spectrometry to characterize modified oligonucleotide impurities and found phosphorothioate oligonucleotide, phosphate diester oligonucleotide and dithionite as impurities in the oligonucleotides. He also demonstrated that the strong anion exchange-ion pair reversed phase two-dimensional liquid chromatography must be an advanced method for determining impurities from oligonucleotides.

3.2.5. Analysis of Therapeutic Antibodies

Hydrophilic interaction liquid chromatography (HILIC) is a type of modern and modified liquid chromatography widely used for separating compounds with a wide range of polarities in conjunction with reversed-phase liquid chromatography (RPLC). Monoclonal antibodies (mAbs) are the main therapeutic class to treat and eliminate tumors, infectious diseases, and inflammatory and autoimmune disorders. D. R. Stoll et al., 2018 [46] studied the development of comprehensive online two-dimensional liquid-chromatography–mass spectrometry using hydrophilic interaction and reversed-phase separation for rapid and deep profiling of therapeutic antibodies. They demonstrated that this method has highly selective separation of various glycoforms of glycosylated monoclonal antibodies like cetuximab compared to other types of two-dimensional liquid chromatography.

3.2.6. Analysis of Impurities in Biopharmaceuticals

Reversed-phase liquid chromatography-reversed phase liquid chromatography is an advanced and versatile two-dimensional liquid chromatography for separating and analyzing compounds with a wide range of hydrophobicity. Impurities are foreign materials that are not the natural compound of interest and can be aided either by accident or generated during collection, manufacture etc. Patrik Petersson et al., 2016 [38] used multiple heart-cutting two-dimensional liquid chromatography-mass spectrometry towards real-time determination of related impurities of bio-pharmaceuticals in salt-based separation methods and analyzed the three days exposed 1mg/ml bovine insulin in sodium phosphate pH 7 at 50°C and found inorganic ions, formic acid, and deamidated insulin etc.

3.3. Metabolite Analysis

3.3.1. Flavonoid O-glycosides

Hydrophilic interaction liquid chromatography (HILIC) is a modern and versatile liquid chromatography widely used for separating compounds with a wide range of polarities in conjunction with reversed-phase liquid chromatography (RPLC). This can also separate glycosidic components like quinochalcone C-glycosides, saponins, anthocyanins and steviol glycosides [83-86]. Chang-Liang Yao et al., 2017 [47] studied an enhanced targeted identification strategy for the selective identification of flavonoid O-glycosides from Carthamus tinctorius by integrating offline two-dimensional liquid chromatography/linear ion-trap-Orbitrap mass spectrometry, high-resolution diagnostic product ions/neutral loss filtering and liquid chromatography-solid phase extraction-nuclear magnetic resonance and identified 6-methoxy kaempferol, isocarthamidin, cantharidin, kaempferol and 6-hydroxy kaempferol as glycosides and demonstrated this method as a convenience for separation and identification of flavonoid O-glycosides.

3.3.2. Furanocoumarins

Reversed-phase liquid chromatography-reversed phase liquid chromatography is an advanced and versatile two-dimensional liquid chromatography for separating and analyzing compounds with a wide range of hydrophobicity. Grapefruit juice consumption can affect the metabolism of various drugs, primarily by inhibiting CYP3A4 by grapefruit juice containing furanocoumarin components [87]. Furanocoumarins may cause photo-sensitivity to the skin. The effect can enhance ultraviolet treatments for skin diseases like psoriasis and vitiligo [88]. Daniel W. Cook et al., 2017 studied a comparison of multivariate curve resolution strategies in quantitative liquid chromatography-liquid chromatography for application to the quantification of furanocoumarins in apiaceous vegetables and analyzed parsley, parsnips, and celery vegetable and found Sphondin, 6,7-dihydroxybergamottin, isoimperatorin, oxypeucedanin, Isopimpinellin and imperatorin, psoralen, angelicin, 8-methoxy psoralen and 5-methoxy psoralen furanocoumarins. He also demonstrated that this method could be a promising approach for quantitative analysis of different components of furanocoumarins.

3.3.3. Plasma Lipids

Normal phase liquid chromatography (NPLC)- reversed-phase liquid chromatography (RPLC) two-dimensional liquid chromatography is an advanced and widely used versatile mode of high-performance liquid chromatography for the analysis of various compounds in different food materials with more polarity to less polarity. There is a relationship between lipid metabolism and stroke. Upregulation of phospholipases, especially phospholipase A2, is responsible for the increase of lysophospholipids and free fatty acids (FFA), thus leading to ischemia-initiated events [89-91]. Li Yang et al., 2017 [89] studied lipidomic analysis of plasma in patients with lacunar infarction using normal phase/reversed-phase two-dimensional liquid chromatography–quadrupole time of flight mass spectrometry. They found phosphatidylcholine, lysophosphatidylcholine, phosphatidylethanolamine, lysophosphatidylethanolamine, phosphatidylglycerol, lysophosphatidyl-glycerol, phosphatidylinositol, phosphatidylserine, sphingomyelin, ceramide, galactosylceramide, glucosylceramide, lactosylceramide, monoacylglycerol, diacylglycerol, triacylglycerol lipid profiles. He demonstrated that this method could be a promising approach for profiling lipids from patient plasma for disease diagnosis.

3.3.4. Vitamin D Metabolites

Reversed-phase liquid chromatography-reversed phase liquid chromatography is an advanced and versatile two-dimensional liquid chromatography for separating and analyzing compounds with a wide range of hydrophobicity, like vitamin D3. Vitamin D is essential for our body to maintain calcium homeostasis and prevent rickets and osteomalacia [92]. Mena-Bravo et al., 2016 [93] used two-dimensional liquid chromatography coupled with tandem mass spectrometry for vitamin D metabolite profiling, including the C3-epimer-25-monohydroxy vitamin D3 and found recovery factor for dihydroxy metabolites 100%, monohydroxy metabolite 95.2%, but for vitamin D2 and D3 59.1% and 60.2% respectively.

3.3.5. Proteinogenic Amino Acids

Reversed phase-chiral two-dimensional liquid chromatography is an advanced and versatile liquid chromatography for separating and analyzing compounds with a wide range of polarity and chirality. The amino acid has the property of both polarity and chirality. Amino acid is a basic protein unit with different roles in our body, like metabolism, maintaining our body structure, etc. Ulrich Woiwode et al., 2018 [50] used an enantioselective multiple heartcut two-dimensional ultra-high-performance liquid chromatography method with a core-shell chiral stationary phase in the second dimension to analyze all proteinogenic amino acids in a single run. They found different amino acids in bacitracin (His, Arg, Ser, Asp, Glu, Thr, Ala, Pro, Val, Aile, Phe, Ile, Leu, Cys, Orn and Lys) and gramicidin (Gly, Ala, Val, Trp, Phe, Ile, Leu, Tyr). He demonstrated this method as a convenient approach for separating and analyzing chiral substances with a wide range of polarity.

3.3.6. Steroids and Sulphonamide Metabolite in Urine

Reversed-phase liquid chromatography-reversed phase liquid chromatography is an advanced and versatile two-dimensional liquid chromatography for separating and analyzing compounds with a wide range of lipophilicity. Metabolites are the compound generated after the metabolism of the ingested substances. The body's primary function is to eliminate the ingested drug by producing its metabolites. M. H. Blokland et al., 2018 [94] studied multiclass screening in urine by comprehensive two-dimensional liquid chromatography time of flight mass spectrometry for residues of sulfonamides, beta-agonists and steroids and found that sulfonamides and β-agonists are fully partly separated and steroids are fully separated. He demonstrated that this method could separate and analyze urine samples for beta-agonist, sulphonamides and steroids.

3.4. Environmental Analysis

3.4.1. Emerging Contaminants

Reversed phase-reversed phase is an advanced and versatile two-dimensional liquid chromatography for separating and analyzing compounds with a wide range of lipophilicity. Contaminants are unwanted materials or substances in natural pure compounds or substances. Juri Leonhardt et al., 2016 [95] studied a new method for determining peak distribution across a two-dimensional separation space to identify optimal column combinations by using the emerging contaminant and developed a method for selective and high-resolution analysis of the contaminants.

3.4.2. Polycyclic Aromatics

Comprehensive reversed phase-reversed phase two-dimensional liquid chromatography is an advanced and versatile method for separating and analyzing compounds with a wide range of lipophilicity. Rune Graesboll et al., 2017 [96] studied optimizing gradient conditions in online comprehensive two-dimensional reversed-phase liquid chromatography using a mixture of 25 polycyclic aromatic compounds using the linear solvent strength model. They demonstrated that the comprehensive two-dimensional reversed phase-reversed phase liquid chromatography approach is expected to efficiently separate complex mixtures of polycyclic aromatic compounds in the environment.

3.4.3. Waste Water Effluent

Reversed phase-reversed phase is an advanced and versatile two-dimensional liquid chromatography for separating and analyzing compounds with a wide range of polarities. Environmental analysis is a complex task due to human activities and the increasing number of compounds in the environment. Wastewater is a type of water already used for any purpose and discarded in the environment. Xiyu Ouyang et al., 2016 [97] studied the rapid screening of acetylcholinesterase (AchE) inhibitors by effect-directed analysis using LC ×

LC fractionation, high throughput in vitro assay, and parallel identification by the time of flight mass spectrometry and analyzed wastewater treatment plant effluent and found tiapride, amisulpride, and lamotrigine AchE inhibitors. It is demonstrated that this method could be used to analyze wastewater for AchE inhibitors with greater peak capacity (0.937) and excellent orthogonality (0.945).

3.5. Natural Medicine Analysis

3.5.1. Resolution of Complex Mixture of Senna Glycosides

Gel permeation chromatography is a versatile liquid chromatography widely used for the separation and high resolution of a complex mixture based on the molecular size and weight as well as lipophilicity in conjunction with reversed-phase liquid chromatography. Complex mixtures are those mixtures containing large no. of single components. F. Erni et al., 1978 [9] used a two-dimensional column liquid chromatographic technique for the resolution of complex mixtures of senna glycosides and found two separate sennoside A (minor) and sennoside B (major) senna glycoside. He demonstrated that two-dimensional gel permeation-reversed phase liquid chromatography could be a selective approach for resolving a complex mixture of substances or drugs.

3.5.2. Analysis of Brown Seaweeds

Hydrophilic interaction liquid chromatography (HILIC) is a modern and versatile liquid chromatography widely used for separating compounds with a wide range of polarities in conjunction with reversed-phase liquid chromatography (RPLC). Brown seaweeds contain phlorotannins which are polyphenolic compounds. These compound types possess several bioactivities like antioxidant, anti-inflammatory, antibacterial, anti-diabetic and anti-adipogenic [98-106]. Lidia Montero et al., 2016 [107] studied the antiproliferative activity and chemical characterization by comprehensive two-dimensional liquid chromatography coupled to mass spectrometry of phlorotannins from the brown macroalga Sargassum muticum collected on North-Atlantic coasts and found different phlorotannins of Sargassum muticum of other territories such as trifuhalol, dihydroxytetrafuhalol, pentafuhalol, trihydroxyhexafuhalol etc. for sample from Portugal and trifuhalol, dihydroxytetrafuhalol, pentafuhalol, tetrafuhalol, hydroxy-pentafuhalol etc. for example from Norway. He demonstrated that this method could separate and analyze polyphenols, including phlorotannins.

3.5.3. Analysis of Decoction

Comprehensive two-dimensional reversed phase-reversed phase liquid chromatography is an advanced and versatile mode for separating and analyzing compounds with a wide range of polarities. Gegen-Qinlian decoction is a natural and traditional Chinese medicine widely utilized for treating diarrhoea and diabetes [108]. Xue Qiao et al., 2016 [109] studied a chemical profiling solution for Chinese medicine formulas using comprehensive and loop-based multiple heart-cutting two-dimensional liquid chromatography coupled with quadrupole time-of-flight mass spectrometry and found 125 characterized compounds in the decoction. He concluded that this could be an effective method for analyzing a complex mixture of chemicals like traditional Chinese medicine formulations.

3.5.4. Analysis of Flavonoids and Iridoid Glycosides

Comprehensive two-dimensional reversed-phase liquid chromatography is a modern and versatile type of liquid chromatography for separating and analyzing compounds with a wide range of lipophilicity. Chinese herbal medicine is nowadays gaining increased attention in healthcare. Duxin Li et al., 2014 [110] used a comprehensive two-dimensional liquid chromatography-tandem diode array detector (DAD) and accurate mass QTOF-MS for the analysis of flavonoids and iridoid glycosides in *Hedyotis diffusa* and found five flavonol glycosides, four acylated flavonol glycosides, three iridoid glycosides and several novel flavonoids. This method demonstrated that it provides effective herbal medicine analysis.

3.5.5. Analysis of Chemical Constituents in Ginkgo biloba

Hydrophilic interaction liquid chromatography (HILIC) is a modern and versatile liquid chromatography widely used for separating compounds with a wide range of polarities by coupling with reversed-phase liquid chromatography (RPLC). Ginkgo biloba extract is the most frequently used traditional Chinese medicine worldwide and is prepared using the leaves of Ginkgo biloba L. It is widely used as a medicine or a dietary supplement to treat intermittent claudication, neurodegenerative diseases and tinnitus [111-113]. Shuai Ji et al., 2017 [114] studied the separation and characterization of chemical constituents in Ginkgo biloba extract by offline hydrophilic interaction × reversed-phase two-dimensional liquid chromatography coupled with quadrupole-time of flight mass spectrometry. They found a structure of 104 characterized compounds comprising 74 flavonoid glycosides, 14 free flavonoids, 9 terpene lactones, 5 phenolic acids, and 2 other combinations.

This method demonstrated that it could improve the technique for separating and characterizing natural products like Ginkgo biloba extract or other complex natural medicines.

3.5.6. Analysis of Saponins in Sinensis Fruits

Comprehensive two-dimensional reversed phase-reversed phase liquid chromatography is an advanced and versatile mode of liquid chromatography for separating and analyzing compounds with a wide range of polarities like saponins. Gleditsia sinensis Lam. is a tree native to China and its fruits are widely used as the Chinese herbal medicine for treating apoplexy. More than 5% (w/w) of triterpenoid saponins are found in these fruits. These saponins are used as antitumors and antiallergic agents [115-118]. Shuang Wang et al., 2016 [117] studied the separation and characterization of triterpenoid saponins in Gleditsia sinensis by comprehensive two-dimensional liquid chromatography coupled with mass spectrometry and found 72 saponins by characterization. For the first time, 49 compounds and two acetylated saponins were detected from Gleditsia sinensis. He also demonstrated comprehensive two-dimensional reversed phase-reversed phase liquid chromatography coupling C18 and poroshell 120 phenyl-hexyl columns in first and second dimensions as an effective and reliable method for separating and analyzing triterpenoid saponins.

3.6. Pharmaceutical Analysis

3.6.1. Impurities in Cefonicid Sodium

Reversed phase-reversed phase is an advanced and versatile two-dimensional liquid chromatography for separating and analyzing compounds with a wide range of polarities (lipophilicity). Cefonicid Sodium is the broad-spectrum second-generation cephalosporin antibiotic drug used against gram-negative and gram-positive microorganisms. Jian Wanga et al., 2017 [119] studied the separation and characterization of unknown impurities in cefonicid sodium by trap-free two-dimensional liquid chromatography combined with ion trap/time-of-flight mass spectrometry and identified seven structurally characterized impurities in cefonicid sodium.

3.6.2. Antibiotic Residues in Dairy Products

Two-dimensional liquid chromatography is one of the novel and versatile modes for separating and analyzing compounds with different polarities when used by coupling hydrophilic interaction liquid chromatography with reversed-phase liquid chromatography. Antibiotics are drugs used to treat or eradicate infections or diseases caused by bacteria, fungi, spirochetes, mycoplasma, chlamydia, and other pathogenic microorganism. Lian Wang et al., 2016 [120] used novel two-dimensional liquid chromatography–tandem mass spectrometry for the analysis of twenty antibiotics residues in dairy products and found average recoveries of 77.0% (doxycycline)—102.1% (ciprofloxacin) and 76.4% (tetracycline)— 98.9% (ciprofloxacin), and relative standard deviation of 2.8% (kanamycin)—6.9% (tetracycline) and 4.0% (chloramphenicol)—7.8% (tetracycline), respectively for 25 μg/kg for milk powder and milk sample. Average recoveries of the milk powder and milk sample were found to be 82.4% (tetracycline)—104.7% (enrofloxacin) and 79.0% (tetracycline)—105.8% (ciprofloxacin), and relative standard deviation of 3.6% (chloramphenicol)—7.2% (tetracycline) and 4.2% (streptomycin)— 8.8% (doxycycline), respectively for 75 μg/kg. This is a reliable and effective method for analyzing antibiotics in milk and milk products.

3.6.3. Beta-Blockers in Human Plasma

Size exclusion chromatography is a modern and versatile liquid chromatography for separating compounds with a wide range of molecular weight, size and polarity when used in conjunction with reversed-phase liquid chromatography. Beta-blockers are β-adrenoceptor antagonists used for treating acute coronary syndrome or heart failure, like hypertension, cardiac arrhythmias, angina pectoris, and for cardioprotection following myocardial infarction [121]. Virginia M. F. Goncalves et al., 2017 [122] studied the quantification of propranolol and propranolol in human plasma using two-dimensional liquid chromatography (2D-LC) and found 5 ng/ml lower limit of quantification (LLOQ) 5-200 ng/ml and R2 value 0.9995. He demonstrated this method as accurate, precise, reliable and robust for the analysis of beta-blocker such as alpranolol and propranolol.

3.6.4. Desonide Cream

Reversed phase-reversed phase two-dimensional liquid chromatography is an advanced and versatile liquid chromatography for separating and analyzing compounds with a wide range of polarities (lipophilicity). Desonide is a synthetic corticosteroid widely used topically for treating inflammation caused by allergic reactions and psoriasis [123]. Jian Wang et al., 2018 [124] conducted a study to develop a novel HPLC method for determining the impurities in desonide cream and characterization of its impurities by 2D LC-IT-TOF MS and found structures of four degradation impurities. He also demonstrated this novel HPLC method as simple and reliable for the quality control of desonide cream.

3.6.5. Vancomycin in Human Plasma

Reversed phase-reversed phase two-dimensional liquid chromatography is a novel and versatile liquid chromatography for separating and analyzing compounds with a wide range of polarities. Vancomycin is a glycopeptide antibiotic having solid bactericidal activity against Gram-positive bacteria. Yanghao Sheng et al., 2017 [125] used a cost-effective two-dimensional liquid chromatography system for high throughput determination of vancomycin in human plasma and found 0.20 μg/ml lower limit of quantification (LLOQ) for vancomycin. He demonstrated this method as rapid and automated with low-cost, high selectivity and sensitivity for quantifying vancomycin in human plasma.

3.6.6. Vitamins

Hydrophilic interaction liquid chromatography (HILIC) is a modern and versatile liquid chromatography widely used for separating compounds with large polarities in conjunction with reversed-phase liquid chromatography (RPLC). Multivitamin formulations are used as a nutritional supplement and are recommended for malnutrition, some diseases, intense alcohol abuse, and heavy smoking. Hydrophilic interaction chromatography (HILIC) separates and analyses water-soluble vitamins, while reversed-phase liquid chromatography (RPLC) separates and analyzes fat-soluble vitamins. Stefanie Baurer et al., 2018 [126] separated water and fat-soluble vitamins by selective comprehensive HILIC × RPLC (high-resolution sampling) and active solvent modulation. They investigated water-soluble vitamins (vitamins B1, B2, B3, B6, B9, B12 and vitamin C) and fat-soluble (vitamin A palmitate, vitamin D3, vitamin E and vitamin K). He demonstrated that this method can be used for the simultaneous separation and analysis of fat and water-soluble vitamins and can also be used for the quantitative analysis of multivitamins supplements.

3.7. Miscellaneous Applications Analysis

3.7.1. Tobacco Product

Multiple heart-cutting two-dimensional liquid chromatography is a modern and versatile type of liquid chromatography for separating and analyzing the different types of compounds with different polarities. The reversed-phase liquid chromatography is a method to separate non-polar compounds and it becomes a versatile method in conjunction with pentafluorophenyl stationary phase for the separation of ochratoxin A. Aflatoxin and ochratoxin A are most toxin components found as contaminants in agricultural products due to it's hepatotoxic, carcinogenic, teratogenic effect [127]. Dawei Qi et al., 2017 [128] used multiple heart-cutting two-dimensional liquid chromatography coupled with quadrupole-orbitrap high-resolution mass spectrometry to simultaneously determine aflatoxin and ochratoxin A in snus, a smokeless tobacco product. They separated and identified aflatoxin G1, G2, B1, B2 and ochratoxin A. He demonstrated that this method has good repeatability, recovery and would be a rapid and reliable approach for analyzing aflatoxin and ochratoxin A.

3.7.2. Metaproteomes in Soil

Strong cation exchange (SCX) chromatography is an advanced and novel liquid chromatography that can be utilized for the separation and analysis of a wide range of compounds with cationic and lipophilic components in conjunction with reversed-phase liquid chromatography as a two-dimensional liquid chromatography. Soil metaproteomics is the essential component required for maintaining a suitable soil environment. Stephen J. Callister et al., 2018 [129] carried out a study to address the challenge of soil metaproteome complexity by improving metaproteome depth of coverage through two-dimensional liquid chromatography. They found that the offline approach provides greater soil metaproteome depth of coverage than the online two-dimensional-mass-mass/mass spectrometry approach. However, the online method provides greater soil metaproteome depth of coverage than conventional one-dimensional liquid chromatography.

3.7.3. Peptides

Size exclusion chromatography (SEC)-capillary zone electrophoresis (CZE) is a novel and versatile type of two-dimensional liquid chromatography mainly used for the separation and analysis of compounds based on molecular weight, size and combinations carrying different charges, while the reversed-phase (RPLC)-capillary zone electrophoresis (CZE) two-dimensional liquid chromatography can be used for the separation and analysis of compounds with lipophilicity and compounds carrying an extensive range of different charges. John P. Larmann Jr. et al., 1993 [130] used two-dimensional separations of peptides and proteins by comprehensive liquid chromatography-capillary electrophoresis in the modes such as SEC-CZE, RPLC-CZE and RPLC-fast CZE. They found these three modes of two-dimensional liquid chromatography with greater resolving power than conventional 1D liquid chromatography for separating and analyzing peptides from complex biological mixtures.

3.7.4. Proteins

Strong anion exchange (SAX) chromatography is an advanced and novel liquid chromatography used for the separation and analysis of a wide range of compounds with anionic and lipophilic components in conjunction with reversed-phase as a two-dimensional liquid chromatography. Proteins are the essential components of our body and can be found in plasma, muscles, tissues etc. Zhi Huang et al., 2016 [131] used an array-based online two-dimensional liquid chromatography system applied to the effective depletion of high-abundance proteins in human plasma and found a total of 1332 proteins with a compelling lack of 84 high-abundance proteins. He also demonstrated that this method has a great prospect of high-throughput separation of complex protein samples.

3.7.5. Surfactants

Hydrophilic interaction liquid chromatography (HILIC)- reversed-phase liquid chromatography (RPLC) is a modern and versatile liquid chromatography widely used for separating compounds with an extensive range of polarity. Surfactants are amphiphilic substances having both a polar head and non-polar tail in solution form. Andrea F. G. Gargano et al., 2015 [132] conducted a study to reduce dilution and analysis time in online comprehensive two-dimensional liquid chromatography by active modulation using HILIC-RPLC 2D-LC. They found an analysis time of 40 min, demonstrating that this method can be used for rapid analysis by using active modulation.

3.7.6. Synthetic Polymers

Size exclusion chromatography (SEC)- reversed-phase liquid chromatography (RPLC) is a novel and versatile type of two-dimensional liquid chromatography for separating compounds with a wide range of molecular weight, size and polarity. Polymers consist of compounds with significant heterogeneity. and exhibit distributions in more than one dimension, like molecular weight distribution, functionality type distribution, chemical composition distribution and various branching distributions etc. [133]. Matthias Pursch et al., 2018 [39] studied the evaluation of active solvent modulation to enhance two-dimensional liquid chromatography for target analysis in polymeric matrices using SEC-RPLC two-dimensional liquid chromatography. They analyzed the polymer blend consisting of epoxy novolac and phenol novolac using an analysis time of 20 minutes which demonstrated that this method can be used for rapid separation and analysis of polymeric complex samples.

Conclusion

Two-dimensional liquid chromatography is a novel and versatile hyphenated liquid chromatography that is diversely used to separate and analyze a very complex mixture of components of interest. The advantage of 2D-LC is that a challenging sample combination can be analyzed rapidly with high sensitivity and high peak capacity by a single injection into the first-dimension column followed by the second-dimension column. 2D-LC has some limitations, such as solvent incompatibility in those types of 2D-LC which have columns with different mechanisms for separation in both first and second dimensions like two-dimensional normal phase liquid chromatography (NPLC)-reversed-phase liquid chromatography (RPLC) with a challenging process for method development and it is the costly instrument. But it has diverse applications in different fields, such as food analysis, pharmaceutical and biopharmaceutics analysis, environmental analysis, metabolites and natural medicines analysis. Thus we can conclude that two-dimensional liquid chromatography can be applied in the separation and analysis of compounds with a wide range of polarity, ionization, molecular weight, size and chirality with high resolution, selectivity, sensitivity, reproducibility, orthogonality and high peak capacity.

References

[1] François, I., A. De Villiers, B. Tienpont, F. David, P. Sandra, Comprehensive two-dimensional liquid chromatography applying two parallel columns in the second dimension, *Journal of Chromatography A*, 1178 (2008) 33-42.

[2] Schoenmakers, P., P. Marriott, J. Beens, Nomenclature and conventions in comprehensive multidimensional chromatography, *LCGC Europe*, 25 (2003) 1-4.

[3] Apffel, J., T. V. Alfredson, R. E. Majors, Automated on-line multi-dimensional high-performance liquid chromatographic techniques for the clean-up and analysis of water-soluble samples, *Journal of Chromatography A*, 206 (1981) 43-57.

[4] Guillarme, D., J. L. Veuthey, Theory and Practice of UHPLC and UHPLC–MS, *Handbook of Advanced Chromatography/Mass Spectrometry Techniques*, Elsevier2017, pp. 1-38.

[5] Stoll, D. R., P. W. Carr, Two-dimensional liquid chromatography: a state of the art tutorial, *Analytical chemistry*, 89 (2017) 519-531.

[6] Stoll, D. R., X. Li, X. Wang, P. W. Carr, S. E. Porter, S. C. Rutan, Fast, comprehensive two-dimensional liquid chromatography, *Journal of Chromatography A*, 1168 (2007) 3-43.

[7] Groeneveld, G., M. N. Dunkle, M. Rinken, A. F. Gargano, A. de Niet, M. Pursch, E. P. Mes, P. J. Schoenmakers, Characterization of complex polyether polyols using comprehensive two-dimensional liquid chromatography hyphenated to high-resolution mass spectrometry, *Journal of Chromatography A*, 1569 (2018) 128-138.

[8] Pirok, B. W., D. R. Stoll, P. J. Schoenmakers, Recent developments in two-dimensional liquid chromatography: fundamental improvements for practical applications, *Analytical Chemistry*, 91 (2018) 240-263.

[9] Erni, F., R. Frei, Two-dimensional column liquid chromatographic technique for resolution of complex mixtures, *Journal of Chromatography A*, 149 (1978) 561-569.

[10] Pirok, B. W., A. F. Gargano, P. J. Schoenmakers, Optimizing separations in online comprehensive two-dimensional liquid chromatography, *Journal of separation science*, 41 (2018) 68-98.

[11] Bedani, F., P. J. Schoenmakers, H. G. Janssen, Theories to support method development in comprehensive two-dimensional liquid chromatography–A review, *Journal of separation science*, 35 (2012) 1697-1711.

[12] Murphy, R. E., M. R. Schure, J. P. Foley, Effect of sampling rate on resolution in comprehensive two-dimensional liquid chromatography, *Analytical Chemistry,* 70 (1998) 1585-1594.

[13] Stoll, D. R., K. Shoykhet, P. Petersson, S. Buckenmaier, Active solvent modulation: a valve-based approach to improve separation compatibility in two-dimensional liquid chromatography, *Analytical chemistry*, 89 (2017) 9260-9267.

[14] Filgueira, M. R., Y. Huang, K. Witt, C. Castells, P. W. Carr, Improving peak capacity in fast online comprehensive two-dimensional liquid chromatography with post-first-dimension flow splitting, *Analytical chemistry,* 83 (2011) 9531-9539.

[15] François, I., A. De Villiers, P. Sandra, Considerations on the possibilities and limitations of comprehensive normal phase–reversed phase liquid chromatography (NPLC× RPLC), *Journal of separation science*, 29 (2006) 492-498.

[16] Guiochon, G. The limits of the separation power of unidimensional column liquid chromatography, *Journal of Chromatography A*, 1126 (2006) 6-49.

[17] Carr, P. W., X. Wang, D. R. Stoll, Effect of pressure, particle size, and time on optimizing performance in liquid chromatography, *Analytical chemistry*, 81 (2009) 5342-5353.

[18] Dsouza, B. R., A. Joseph, S. Das, A. T. Alex, Multiple heart-cutting two-dimensional liquid chromatography: recent developments and applications, *Current Analytical Chemistry*, 17 (2021) 339-354.

[19] Stoll, D. R., E. S. Talus, D. C. Harmes, K. Zhang, Evaluation of detection sensitivity in comprehensive two-dimensional liquid chromatography separations of an active pharmaceutical ingredient and its degradants, *Analytical and bioanalytical chemistry*, 407 (2015) 265-277.

[20] Groskreutz, S. R., M. M. Swenson, L. B. Secor, D. R. Stoll, Selective comprehensive multi-dimensional separation for resolution enhancement in high performance liquid chromatography. Part I: principles and instrumentation, *Journal of Chromatography A,* 1228 (2012) 31-40.

[21] Česla, P., J. Křenková, Fraction transfer process in on-line comprehensive two-dimensional liquid-phase separations, *Journal of Separation Science,* 40 (2017) 109-123.

[22] Egeness, M. J., M. C. Breadmore, E. F. Hilder, R. Shellie, The modulator in comprehensive two-dimensional liquid chromatography, *LC GC Europe*, 29 (2016) 268-276.

[23] Murphy, R. E., M. R. Schure, J. P. Foley, One-and two-dimensional chromatographic analysis of alcohol ethoxylates, *Analytical Chemistry*, 70 (1998) 4353-4360.

[24] Kar, A. *Pharmaceutical Drug Analysis*, second edition ed., New Age International (P) Limited, New Delhi, India, 2001.

[25] Mondello, L., P. Tranchida, V. Stanek, P. Jandera, G. Dugo, P. Dugo, Silver-ion reversed-phase comprehensive two-dimensional liquid chromatography combined with mass spectrometric detection in lipidic food analysis, *Journal of Chromatography A*, 1086 (2005) 91-98.

[26] Guo, K., J. Peng, R. Zhou, L. Li, Ion-pairing reversed-phase liquid chromatography fractionation in combination with isotope labeling reversed-phase liquid chromatography–mass spectrometry for comprehensive metabolome profiling, *Journal of chromatography A*, 1218 (2011) 3689-3694.

[27] Dugo, P., N. Fawzy, F. Cichello, F. Cacciola, P. Donato, L. Mondello, Stop-flow comprehensive two-dimensional liquid chromatography combined with mass spectrometric detection for phospholipid analysis, *Journal of Chromatography A*, 1278 (2013) 46-53.

[28] Van Nederkassel, A. M., A. Aerts, A. Dierick, D. Massart, Y. Vander Heyden, Fast separations on monolithic silica columns: method transfer, robustness and column

ageing for some case studies, *Journal of pharmaceutical and biomedical analysis,* 32 (2003) 233-249.

[29] Eeltink, S., G. Desmet, G. Vivó-Truyols, G. P. Rozing, P. J. Schoenmakers, W. T. Kok, Performance limits of monolithic and packed capillary columns in high-performance liquid chromatography and capillary electrochromatography, *Journal of Chromatography A*, 1104 (2006) 256-262.

[30] Nazario, C. E., M. R. Silva, M. S. Franco, F. M. Lanças, Evolution in miniaturized column liquid chromatography instrumentation and applications: An overview, *Journal of Chromatography a*, 1421 (2015) 18-37.

[31] Martin, M., G. Blu, C. Eon, G. Guiochon, The use of syringe-type pumps in liquid chromatography in order to achieve a constant flow-rate, *Journal of Chromatography A,* 112 (1975) 399-414.

[32] Simpson, C. High-performance liquid chromatography. Injection techniques in liquid chromatography, *Proceedings of the Analytical Division of the Chemical Society*, *Royal Society of Chemistry*, 1979, pp. 222-228.

[33] Paul, C., F. Steiner, M. W. Dong, HPLC autosamplers: Perspectives, principles, and practices, *LC-GC North Am,* 37 (2019) 514-529.

[34] Gilar, M., A. E. Daly, M. Kele, U. D. Neue, J. C. Gebler, Implications of column peak capacity on the separation of complex peptide mixtures in single-and two-dimensional high-performance liquid chromatography, *Journal of chromatography A*, 1061 (2004) 183-192.

[35] Schoenmakers, P. J., G. Vivó-Truyols, W. M. Decrop, A protocol for designing comprehensive two-dimensional liquid chromatography separation systems, *Journal of Chromatography A*, 1120 (2006) 282-290.

[36] Bathke, A., D. Klemm, C. Gstottner, C. Bell, R. Kopf, Rapid Online Reduction and Characterization of Protein Modifications Using Fully Automated Two-Dimensional High Performance Liquid Chromatography-Mass Spectrometry, *LC-GC North America*, 36 (2018) 18-29.

[37] Li, Y., C. Stella, L. Zheng, C. Bechtel, J. Gruenhagen, F. Jacobson, C. D. Medley, Investigation of low recovery in the free drug assay for antibody drug conjugates by size exclusion—reversed phase two dimensional-liquid chromatography, *Journal of Chromatography B,* 1032 (2016) 112-118.

[38] Petersson, P., K. Haselmann, S. Buckenmaier, Multiple heart-cutting two dimensional liquid chromatography mass spectrometry: Towards real time determination of related impurities of bio-pharmaceuticals in salt based separation methods, *Journal of Chromatography A*, 1468 (2016) 95-101.

[39] Pursch, M., A. Wegener, S. Buckenmaier, Evaluation of active solvent modulation to enhance two-dimensional liquid chromatography for target analysis in polymeric matrices, *Journal of Chromatography A,* 1562 (2018) 78-86.

[40] Yang, P., W. Gao, J. E. Shulman, Y. Chen, Separation and identification of polymeric dispersants in detergents by two-dimensional liquid chromatography, *Journal of Chromatography A*, 1566 (2018) 111-117.

[41] Roussis, S. G., I. Cedillo, C. Rentel, Two-dimensional liquid chromatography-mass spectrometry for the characterization of modified oligonucleotide impurities, *Analytical biochemistry*, 556 (2018) 45-52.

[42] Cacciola, F., D. Giuffrida, M. Utczas, D. Mangraviti, M. Beccaria, P. Donato, I. Bonaccorsi, P. Dugo, L. Mondello, Analysis of the carotenoid composition and stability in various overripe fruits by comprehensive two-dimensional liquid chromatography, *Lc Gc Europe*, 29 (2016) 252-257.

[43] Cacciola, F., D. Giuffrida, M. Utczas, D. Mangraviti, P. Dugo, D. Menchaca, E. Murillo, L. Mondello, Application of comprehensive two-dimensional liquid chromatography for carotenoid analysis in red mamey (*Pouteria sapote*) fruit, *Food Analytical Methods*, 9 (2016) 2335-2341.

[44] Brazdauskas, T., L. Montero, P. R. Venskutonis, E. Ibañez, M. Herrero, Downstream valorization and comprehensive two-dimensional liquid chromatography-based chemical characterization of bioactives from black chokeberries (*Aronia melanocarpa*) pomace, *Journal of Chromatography A,* 1468 (2016) 126-135.

[45] Williams, A., E. K. Read, C. D. Agarabi, S. Lute, K. A. Brorson, Automated 2D-HPLC method for characterization of protein aggregation with in-line fraction collection device, *Journal of Chromatography B*, 1046 (2017) 122-130.

[46] Stoll, D. R., D. C. Harmes, G. O. Staples, O. G. Potter, C. T. Dammann, D. Guillarme, A. Beck, Development of comprehensive online two-dimensional liquid chromatography/mass spectrometry using hydrophilic interaction and reversed-phase separations for rapid and deep profiling of therapeutic antibodies, *Analytical chemistry,* 90 (2018) 5923-5929.

[47] Yao, C. L., W. Z. Yang, W. Si, Y. Shen, N. X. Zhang, H. L. Chen, H. Q. Pan, M. Yang, W. Y. Wu, D. A. Guo, An enhanced targeted identification strategy for the selective identification of flavonoid O-glycosides from Carthamus tinctorius by integrating offline two-dimensional liquid chromatography/linear ion-trap-Orbitrap mass spectrometry, high-resolution diagnostic product ions/neutral loss filtering and liquid chromatography-solid phase extraction-nuclear magnetic resonance, *Journal of Chromatography a,* 1491 (2017) 87-97.

[48] Berkecz, R., F. Tömösi, T. Körmöczi, V. Szegedi, J. Horváth, T. Janáky, Comprehensive phospholipid and sphingomyelin profiling of different brain regions in mouse model of anxiety disorder using online two-dimensional (HILIC/RP)-LC/MS method, *Journal of pharmaceutical and biomedical analysis*, 149 (2018) 308-317.

[49] Barhate, C. L., E. L. Regalado, N. D. Contrella, J. Lee, J. Jo, A. A. Makarov, D. W. Armstrong, C. J. Welch, Ultrafast chiral chromatography as the second dimension in two-dimensional liquid chromatography experiments, *Analytical chemistry,* 89 (2017) 3545-3553.

[50] Woiwode, U., S. Neubauer, W. Lindner, S. Buckenmaier, M. Lämmerhofer, Enantioselective multiple heartcut two-dimensional ultra-high-performance liquid chromatography method with a Coreshell chiral stationary phase in the second dimension for analysis of all proteinogenic amino acids in a single run, *Journal of Chromatography A,* 1562 (2018) 69-77.

[51] Navarro-Reig, M., J. Jaumot, T. A. van Beek, G. Vivó-Truyols, R. Tauler, Chemometric analysis of comprehensive LC×LC-MS data: resolution of triacylglycerol structural isomers in corn oil, *Talanta,* 160 (2016) 624-635.

[52] Vonk, R. J., A. F. Gargano, E. Davydova, H. L. Dekker, S. Eeltink, L. J. De Koning, P. J. Schoenmakers, Comprehensive two-dimensional liquid chromatography with stationary-phase-assisted modulation coupled to high-resolution mass spectrometry applied to proteome analysis of *Saccharomyces cerevisiae*, *Analytical chemistry,* 87 (2015) 5387-5394.

[53] Mondello, L. *Comprehensive chromatography in combination with mass spectrometry,* John Wiley & Sons2011.

[54] Dugo, P., T. Kumm, F. Cacciola, G. Dugo, L. Mondello, Multidimensional liquid chromatographic separations applied to the analysis of food samples, *Journal of Liquid Chromatography & Related Technologies®*, 31 (2008) 1758-1807.

[55] Swartz, M. HPLC detectors: a brief review, *Journal of Liquid Chromatography & Related Technologies,* 33 (2010) 1130-1150.

[56] Harvey, D. *Modern analytical chemistry*, McGraw-Hill New York2000.

[57] Pavia, D. L., G. M. Lampman, G. S. Kriz, J. A. Vyvyan, *Introduction to spectroscopy*, Cengage learning 2014.

[58] Mourey, T. H., L. E. Oppenheimer, Principles of operation of an evaporative light-scattering detector for liquid chromatography, *Analytical Chemistry*, 56 (1984) 2427-2434.

[59] Blau, K., G. S. King, *Handbook of derivatives for chromatography*, Heyden1978.

[60] Leach, G., G. Oliveira, R. Morais, Production of a carotenoid-rich product by alginate entrapment and fluid-bed drying of *Dunaliella salina*, *Journal of the Science of Food and Agriculture*, 76 (1998) 298-302.

[61] de Quirós, A. R. B., H. S. Costa, Analysis of carotenoids in vegetable and plasma samples: A review, *Journal of Food Composition and Analysis*, 19 (2006) 97-111.

[62] Beutner, S., B. Bloedorn, S. Frixel, I. Hernández Blanco, T. Hoffmann, H. D. Martin, B. Mayer, P. Noack, C. Ruck, M. Schmidt, Quantitative assessment of antioxidant properties of natural colorants and phytochemicals: carotenoids, flavonoids, phenols and indigoids. The role of β-carotene in antioxidant functions, *Journal of the Science of Food and Agriculture*, 81 (2001) 559-568.

[63] Arab, L., S. Steck, Lycopene and cardiovascular disease, *The American journal of clinical nutrition*, 71 (2000) 1691S-1695S.

[64] Rao, A. V., L. G. Rao, Carotenoids and human health, *Pharmacological research*, 55 (2007) 207-216.

[65] Omoni, A. O., R. E. Aluko, The anti-carcinogenic and anti-atherogenic effects of lycopene: a review, *Trends in Food Science & Technology*, 16 (2005) 344-350.

[66] Nishino, H., H. Tokuda, Y. Satomi, M. Masuda, P. Bu, M. Onozuka, S. Yamaguchi, Y. Okuda, J. Takayasu, J. Tsuruta, *Cancer prevention by carotenoids, Pure and Applied Chemistry*, 71 (1999) 2273-2278.

[67] Snodderly, D. M. Evidence for protection against age-related macular degeneration by carotenoids and antioxidant vitamins, *The American journal of clinical nutrition*, 62 (1995) 1448S-1461S.

[68] Dugo, P., M. Herrero, D. Giuffrida, T. Kumm, G. Dugo, L. Mondello, Application of comprehensive two-dimensional liquid chromatography to elucidate the native carotenoid composition in red orange essential oil, *Journal of Agricultural and Food Chemistry*, 56 (2008) 3478-3485.

[69] Montenegro-Landívar, M. F., P. Tapia-Quirós, X. Vecino, M. Reig, C. Valderrama, M. Granados, J. L. Cortina, J. Saurina, Polyphenols and their potential role to fight viral diseases: An overview, *Science of the Total Environment,* 801 (2021) 149719.

[70] Sommella, E., O. H. Ismail, F. Pagano, G. Pepe, C. Ostacolo, G. Mazzoccanti, M. Russo, E. Novellino, F. Gasparrini, P. Campiglia, Development of an improved online comprehensive hydrophilic interaction chromatography× reversed-phase ultra-high-pressure liquid chromatography platform for complex multiclass polyphenolic sample analysis, *Journal of separation science*, 40 (2017) 2188-2197.

[71] Carlson, J. L., J. M. Erickson, B. B. Lloyd, J. L. Slavin, Health effects and sources of prebiotic dietary fiber, *Current developments in nutrition*, 2 (2018) nzy005.

[72] A. Martín-Ortiz, A. I. Ruiz-Matute, M. L. Sanz, F. J. Moreno, M. Herrero, Separation of di-and trisaccharide mixtures by comprehensive two-dimensional liquid chromatography. Application to prebiotic oligosaccharides, *Analytica Chimica Acta*, 1060 (2019) 125-132.

[73] Murphy, P. A., S. Hendrich, C. Landgren, C. M. Bryant, Food mycotoxins: an update, *Journal of food science*, 71 (2006) R51-R65.

[74] Campone, L., S. Rizzo, A. L. Piccinelli, R. Celano, I. Pagano, M. Russo, M. Labra, L. Rastrelli, Determination of mycotoxins in beer by multi heart-cutting two-dimensional liquid chromatography tandem mass spectrometry method, *Food chemistry*, 318 (2020) 126496.

[75] Wu, X., L. Gu, R. L. Prior, S. McKay, Characterization of anthocyanins and proanthocyanidins in some cultivars of Ribes, Aronia, and Sambucus and their antioxidant capacity, *Journal of agricultural and food chemistry*, 52 (2004) 7846-7856.

[76] Nowak, D., M. Gośliński, E. Wojtowicz, Comparative analysis of the antioxidant capacity of selected fruit juices and nectars: Chokeberry juice as a rich source of polyphenols, *International Journal of Food Properties*, 19 (2016) 1317-1324.

[77] Larson, R. A., E. L. Sievers, E. A. Stadtmauer, B. Löwenberg, E. H. Estey, H. Dombret, M. Theobald, D. Voliotis, J. M. Bennett, M. Richie, Final report of the efficacy and safety of gemtuzumab ozogamicin (Mylotarg) in patients with CD33-positive acute myeloid leukemia in first recurrence, *Cancer*, 104 (2005) 1442-1452.

[78] Venkatramani, C., S. R. Huang, M. Al-Sayah, I. Patel, L. Wigman, High-Resolution 2D-LC Analysis of Key Linker Drug Intermediate Used in Antibody Drug Conjugates (ADCs), *Journal of chromatography*, (2017).

[79] D'Atri, V., S. Fekete, A. Beck, M. Lauber, D. Guillarme, Hydrophilic interaction chromatography hyphenated with mass spectrometry: a powerful analytical tool for the comparison of originator and biosimilar therapeutic monoclonal antibodies at the middle-up level of analysis, *Analytical chemistry*, 89 (2017) 2086-2092.

[80] Ouyang, Y., Y. Zeng, Y. Rong, Y. Song, L. Shi, B. Chen, X. Yang, N. Xu, R. J. Linhardt, Z. Zhang, Profiling analysis of low molecular weight heparins by multiple heart-cutting two dimensional chromatography with quadruple time-of-flight mass spectrometry, *Analytical chemistry*, 87 (2015) 8957-8963.

[81] Li, Y., D. Hewitt, Y. K. Lentz, J. A. Ji, T. Y. Zhang, K. Zhang, Characterization and stability study of polysorbate 20 in therapeutic monoclonal antibody formulation by multidimensional ultrahigh-performance liquid chromatography–

charged aerosol detection–mass spectrometry, *Analytical chemistry*, 86 (2014) 5150-5157.

[82] He, Y., O. V. Friese, M. R. Schlittler, Q. Wang, X. Yang, L. A. Bass, M. T. Jones, On-line coupling of size exclusion chromatography with mixed-mode liquid chromatography for comprehensive profiling of biopharmaceutical drug product, *Journal of Chromatography A*, 1262 (2012) 122-129.

[83] Yang, W., W. Si, J. Zhang, M. Yang, H. Pan, J. Wu, S. Qiu, C. Yao, J. Hou, W. Wu, Selective and comprehensive characterization of the quinochalcone C-glycoside homologs in *Carthamus tinctorius* L. by offline comprehensive two-dimensional liquid chromatography/LTQ-Orbitrap MS coupled with versatile data mining strategies, *RSC advances*, 6 (2016) 495-506.

[84] Qiu, S., W. Z. Yang, X. J. Shi, C. L. Yao, M. Yang, X. Liu, B. H. Jiang, W. Y. Wu, D. A. Guo, A green protocol for efficient discovery of novel natural compounds: characterization of new ginsenosides from the stems and leaves of Panax ginseng as a case study, *Analytica chimica acta*, 893 (2015) 65-76.

[85] Jin, H., J. Zhao, W. Zhou, A. Shen, F. Yang, Y. Liu, Z. Guo, X. Zhang, Y. Tao, X. Peng, Preparative separation of a challenging anthocyanin from *Lycium ruthenicum* Murr. by two-dimensional reversed-phase liquid chromatography/hydrophilic interaction chromatography, *RSC advances*, 5 (2015) 62134-62141.

[86] Fu, Q., Z. Guo, X. Zhang, Y. Liu, X. Liang, Comprehensive characterization of Stevia rebaudiana using two-dimensional reversed-phase liquid chromatography/ hydrophilic interaction liquid chromatography, *Journal of separation science,* 35 (2012) 1821-1827.

[87] Guo, L. Q., Y. Yamazoe, Inhibition of cytochrome P450 by furanocoumarins in grapefruit juice and herbal medicines, *Acta Pharmacologica Sinica*, 25 (2004) 129-136.

[88] Kang, A. Y., L. R. Young, C. Dingfelder, S. Peterson, Effects of furanocoumarins from apiaceous vegetables on the catalytic activity of recombinant human cytochrome P-450 1A2, *The protein journal*, 30 (2011) 447-456.

[89] Yang, L., P. Lv, W. Ai, L. Li, S. Shen, H. Nie, Y. Shan, Y. Bai, Y. Huang, H. Liu, Lipidomic analysis of plasma in patients with lacunar infarction using normal-phase/reversed-phase two-dimensional liquid chromatography–quadrupole time-of-flight mass spectrometry, *Analytical and bioanalytical chemistry*, 409 (2017) 3211-3222.

[90] Wang, H. Y. J., C. B. Liu, H. W. Wu, J. S. Kuo, Direct profiling of phospholipids and lysophospholipids in rat brain sections after ischemic stroke, *Rapid Communications in Mass Spectrometry*, 24 (2010) 2057-2064.

[91] Yaemsiri, S., S. Sen, L. F. Tinker, W. R. Robinson, R. W. Evans, W. Rosamond, S. Wasserthiel-Smoller, K. He, Serum fatty acids and incidence of ischemic stroke among postmenopausal women, *Stroke*, 44 (2013) 2710-2717.

[92] Bailey, D., K. Veljkovic, M. Yazdanpanah, K. Adeli, Analytical measurement and clinical relevance of vitamin D3 C3-epimer, *Clinical biochemistry*, 46 (2013) 190-196.

[93] Mena-Bravo, A., F. Priego-Capote, M. L. De Castro, Two-dimensional liquid chromatography coupled to tandem mass spectrometry for vitamin D metabolite

profiling including the C3-epimer-25-monohydroxyvitamin D3, *Journal of Chromatography A,* 1451 (2016) 50-57.

[94] Blokland, M., P. Zoontjes, L. Van Ginkel, M. Van De Schans, S. Sterk, T. Bovee, Multiclass screening in urine by comprehensive two-dimensional liquid chromatography time of flight mass spectrometry for residues of sulphonamides, beta-agonists and steroids, *Food Additives & Contaminants: Part A,* 35 (2018) 1703-1715.

[95] Leonhardt, J., T. Teutenberg, G. Buschmann, O. Gassner, T. C. Schmidt, A new method for the determination of peak distribution across a two-dimensional separation space for the identification of optimal column combinations, *Analytical and Bioanalytical Chemistry*, 408 (2016) 8079-8088.

[96] Græsbøll, R., H. G. Janssen, J. H. Christensen, N. J. Nielsen, Optimizing gradient conditions in online comprehensive two-dimensional reversed-phase liquid chromatography by use of the linear solvent strength model, *Journal of Separation Science*, 40 (2017) 3612-3620.

[97] Ouyang, X., P. E. Leonards, Z. Tousova, J. Slobodnik, J. de Boer, M. H. Lamoree, Rapid screening of acetylcholinesterase inhibitors by effect-directed analysis using LC× LC fractionation, a high throughput in vitro assay, and parallel identification by time of flight mass spectrometry, *Analytical chemistry*, 88 (2016) 2353-2360.

[98] Ragan, M. A Phlorotannins, brown algal polyphenols, *Progress in phycological research*, 4 (1986) 177-241.

[99] Tierney, M. S., T. J. Smyth, M. Hayes, A. Soler-Vila, A. K. Croft, N. Brunton, Influence of pressurised liquid extraction and solid–liquid extraction methods on the phenolic content and antioxidant activities of Irish macroalgae, *International journal of food science & technology*, 48 (2013) 860-869.

[100] Wang, T., R. S. Jónsdóttir, H. Liu, L. Gu, H. G. Kristinsson, S. Raghavan, G. N. Ólafsdóttir, Antioxidant capacities of phlorotannins extracted from the brown algae *Fucus vesiculosus*, *Journal of agricultural and food chemistry*, 60 (2012) 5874-5883.

[101] Anaëlle, T., E. S. Leon, V. Laurent, I. Elena, J. A. Mendiola, C. Stéphane, K. Nelly, M. Luc, S. P. Valérie, Green improved processes to extract bioactive phenolic compounds from brown macroalgae using *Sargassum muticum* as model, *Talanta,* 104 (2013) 44-52.

[102] Lopes, G., C. Sousa, L. R. Silva, E. Pinto, P. B. Andrade, J. Bernardo, T. Mouga, P. Valentão, Can phlorotannins purified extracts constitute a novel pharmacological alternative for microbial infections with associated inflammatory conditions? *PloS one*, 7 (2012) e31145.

[103] Lopes, G., E. Pinto, P. B. Andrade, P. Valentao, Antifungal activity of phlorotannins against dermatophytes and yeasts: approaches to the mechanism of action and influence on *Candida albicans* virulence factor, *PLoS One,* 8 (2013) e72203.

[104] Tanniou, A., L. Vandanjon, M. Incera, E. Serrano Leon, V. Husa, J. Le Grand, J. L. Nicolas, N. Poupart, N. Kervarec, A. Engelen, Assessment of the spatial variability of phenolic contents and associated bioactivities in the invasive alga

Sargassum muticum sampled along its European range from Norway to Portugal, *Journal of Applied Phycology*, 26 (2014) 1215-1230.

[105] Lee, S. H., Y. J. Jeon, Anti-diabetic effects of brown algae derived phlorotannins, marine polyphenols through diverse mechanisms, *Fitoterapia*, 86 (2013) 129-136.

[106] Jung, H. A., H. J. Jung, H. Y. Jeong, H. J. Kwon, M. Y. Ali, J. S. Choi, Phlorotannins isolated from the edible brown alga *Ecklonia stolonifera* exert anti-adipogenic activity on 3T3-L1 adipocytes by downregulating C/EBPα and PPARγ, *Fitoterapia*, 92 (2014) 260-269.

[107] Montero, L., A. P. Sánchez-Camargo, V. García-Cañas, A. Tanniou, V. Stiger-Pouvreau, M. Russo, L. Rastrelli, A. Cifuentes, M. Herrero, E. Ibáñez, Anti-proliferative activity and chemical characterization by comprehensive two-dimensional liquid chromatography coupled to mass spectrometry of phlorotannins from the brown macroalga *Sargassum muticum* collected on North-Atlantic coasts, *Journal of Chromatography A,* 1428 (2016) 115-125.

[108] Miao, W. J., Q. Wang, T. Bo, M. Ye, X. Qiao, W. Z. Yang, C. Xiang, X. Y. Guan, D. A. Guo, Rapid characterization of chemical constituents and rats metabolites of the traditional Chinese patent medicine Gegen-Qinlian-Wan by UHPLC/DAD/ qTOF-MS, *Journal of Pharmaceutical and Biomedical Analysis*, 72 (2013) 99-108.

[109] Qiao, X., Q. Wang, W. Song, Y. Qian, Y. Xiao, R. An, D. A. Guo, M. Ye, A chemical profiling solution for Chinese medicine formulas using comprehensive and loop-based multiple heart-cutting two-dimensional liquid chromatography coupled with quadrupole time-of-flight mass spectrometry, *Journal of Chromatography A*, 1438 (2016) 198-204.

[110] Li, D., O. J. Schmitz, Comprehensive two-dimensional liquid chromatography tandem diode array detector (DAD) and accurate mass QTOF-MS for the analysis of flavonoids and iridoid glycosides in *Hedyotis diffusa*, *Analytical and bioanalytical chemistry*, 407 (2015) 231-240.

[111] Vellas, B., N. Coley, P.-J. Ousset, G. Berrut, J.-F. Dartigues, B. Dubois, H. Grandjean, F. Pasquier, F. Piette, P. Robert, Long-term use of standardised *Ginkgo biloba* extract for the prevention of Alzheimer's disease (GuidAge): a randomised placebo-controlled trial, *The Lancet Neurology*, 11 (2012) 851-859.

[112] Jacoby, D., E. R. Mohler, Drug treatment of intermittent claudication, *Drugs,* 64 (2004) 1657-1670.

[113] Krauss, P., K. Tziridis, S. Buerbank, A. Schilling, H. Schulze, Therapeutic value of *Ginkgo biloba* extract EGb 761® in an animal model (*Meriones unguiculatus*) for noise trauma induced hearing loss and tinnitus, *PloS one*, 11 (2016) e0157574.

[114] Ji, S., D. D. He, T. Y. Wang, J. Han, Z. Li, Y. Du, J. H. Zou, M. Z. Guo, D. Q. Tang, Separation and characterization of chemical constituents in *Ginkgo biloba* extract by off-line hydrophilic interaction× reversed-phase two-dimensional liquid chromatography coupled with quadrupole-time of flight mass spectrometry, *Journal of Pharmaceutical and Biomedical Analysis*, 146 (2017) 68-78.

[115] Pharmacopoeia, C. *Commission. Pharmacopoeia of the People's Republic of China*, Beijing: China Medical Science Press, 2010.

[116] Xia, Y., Z. Gao, Y. Dai, Q. Wang, Analysis of bioactive saponins in Fructus *Gleditsiae abnormalis* and Fructus *Gleditsiae sinensis* by LC-ELSD, *Chromatographia*, 70 (2009) 1361-1366.

[117] Wang, S., Q. Wang, X. Qiao, W. Song, L. Zhong, D.-a. Guo, M. Ye, Separation and characterization of triterpenoid saponins in *Gleditsia sinensis* by comprehensive two-dimensional liquid chromatography coupled with mass spectrometry, *Planta medica*, 82 (2016) 1558-1567.

[118] Dai, Y., Y. P. Chan, L. M. Chu, P. P. H. But, Antiallergic and anti-inflammatory properties of the ethanolic extract from *Gleditsia sinensis*, *Biological and Pharmaceutical Bulletin*, 25 (2002) 1179-1182.

[119] Wang, J., Y. Xu, Y. Zhang, H. Wang, W. Zhong, Separation and characterization of unknown impurities in cefonicid sodium by trap-free two-dimensional liquid chromatography combined with ion trap time-of-flight mass spectrometry, *Rapid Communications in Mass Spectrometry*, 31 (2017) 1541-1550.

[120] Wang, L., B. Yang, X. Zhang, H. Zheng, Novel two-dimensional liquid chromatography–tandem mass spectrometry for the analysis of twenty antibiotics residues in dairy products, *Food Analytical Methods*, 10 (2017) 2001-2010.

[121] De Peuter, O., F. Lussana, R. Peters, H. Buller, P. Kamphuisen, A systematic review of selective and non-selective beta blockers for prevention of vascular events in patients with acute coronary syndrome or heart failure, Database of Abstracts of Reviews of Effects (DARE): *Quality-assessed Reviews* [Internet], (2009).

[122] Gonçalves, V. M., P. Rodrigues, C. Ribeiro, M. E. Tiritan, Quantification of alprenolol and propranolol in human plasma using a two-dimensional liquid chromatography (2D-LC), *Journal of Pharmaceutical and Biomedical Analysis*, 141 (2017) 1-8.

[123] Antonow, M., R. Lorenzoni, G. Barbosa, A. Ourique, P. Gomes, R. Raffin, Development and physicochemical characterization of desonide-loaded nanocapsule suspensions, *Advances in materials science and engineering*, 2016 (2016).

[124] Wang, J., S. Zheng, Y. Xu, H. Hu, M. Shen, L. Tang, Development of a novel HPLC method for the determination of the impurities in desonide cream and characterization of its impurities by 2D LC-IT-TOF MS, *Journal of Pharmaceutical and Biomedical Analysis*, 161 (2018) 399-406.

[125] Sheng, Y., B. Zhou, High-throughput determination of vancomycin in human plasma by a cost-effective system of two-dimensional liquid chromatography, *Journal of Chromatography A,* 1499 (2017) 48-56.

[126] Bäurer, S., W. Guo, S. Polnick, M. Lämmerhofer, Simultaneous separation of water-and fat-soluble vitamins by selective comprehensive HILIC× RPLC (high-resolution sampling) and active solvent modulation, *Chromatographia,* 82 (2019) 167-180.

[127] Loprieno, N. International Agency for Research on Cancer (IARC) monographs on the evaluation of carcinogenic risk of chemicals to man: "relevance of data on mutagenicity," *Mutation research*, 31 (1975) 210.

[128] Qi, D., T. Fei, H. Liu, H. Yao, D. Wu, B. Liu, Development of multiple heart-cutting two-dimensional liquid chromatography coupled to quadrupole-orbitrap high resolution mass spectrometry for simultaneous determination of aflatoxin B1, B2, G1, G2, and ochratoxin A in snus, a smokeless tobacco product, *Journal of agricultural and food chemistry*, 65 (2017) 9923-9929.

[129] Callister, S. J., T. L. Fillmore, C. D. Nicora, J. B. Shaw, S. O. Purvine, D. J. Orton, R. A. White III, R. J. Moore, M. C. Burnet, E. S. Nakayasu, Addressing the challenge of soil metaproteome complexity by improving metaproteome depth of coverage through two-dimensional liquid chromatography, *Soil Biology and Biochemistry*, 125 (2018) 290-299.

[130] Larmann Jr, J. P., A. V. Lemmo, A. W. Moore Jr, J. W. Jorgenson, Two-dimensional separations of peptides and proteins by comprehensive liquid chromatography-capillary electrophoresis, *Electrophoresis,* 14 (1993) 439-447.

[131] Huang, Z., G. Yan, M. Gao, X. Zhang, Array-based online two dimensional liquid chromatography system applied to effective depletion of high-abundance proteins in human plasma, *Analytical chemistry*, 88 (2016) 2440-2445.

[132] Gargano, A. F., M. Duffin, P. Navarro, P. J. Schoenmakers, Reducing dilution and analysis time in online comprehensive two-dimensional liquid chromatography by active modulation, *Analytical chemistry*, 88 (2016) 1785-1793.

[133] Pasch, H., B. Trathnigg, *HPLC of Polymers*, Springer Science & Business Media1999.

Chapter 8

Conclusion

Rohit Bhatia[1,*]
and Bhupinder Kumar[2]
[1]Department of Pharmaceutical Analysis, ISF College of Pharmacy, Ghal Kalan, Moga, Punjab, India
[2]Department of Pharmaceutical Sciences, HNB Garhwal University, Chauras Campus, Srinagar, Garhwal, Uttarakhand, India

Fluid chromatography is a powerful analytical technique used to separate and identify different components of a mixture. Fluid chromatography has been widely used in a variety of fields, including chemistry, biology, environmental science, and pharmaceuticals, and is essential for the characterization and analysis of complex mixtures. Fluid chromatography offered several advantages that make it a valuable tool for separation and analysis in various fields. Here are some of the key advantages:

1. High Separation Efficiency: Fluid chromatography offers high separation efficiency, which means it can separate complex mixtures into individual components with a high degree of accuracy and precision.
2. Flexibility: Fluid chromatography can be used to separate a wide range of sample types, including small and large molecules, ions, and particles.
3. Sensitivity: Fluid chromatography can detect very low concentrations of analytes, which makes it suitable for the analysis of trace amounts of substances.

* Corresponding Author's Email: bhatiarohit5678@gmail.com.

In: Modern Approaches in Fluid Chromatography
Editors: Rohit Bhatia and Bhupinder Kumar
ISBN: 979-8-88697-856-8

4. Speed: Modern fluid chromatography instruments can provide fast separation times, allowing for rapid analysis of large sample volumes.
5. Automation: Fluid chromatography can be fully automated, which reduces the potential for human error and increases the reproducibility of results.
6. High-throughput analysis: Fluid chromatography can be used for high-throughput analysis, enabling the analysis of multiple samples simultaneously.
7. Compatibility: Fluid chromatography is compatible with a wide range of detection methods, including UV-Vis spectroscopy, mass spectrometry, and electrochemical detection.

Recently many advanced versions of fluid chromatography have been introduced with advanced instrumentation and features. This book has described such techniques in detail including their working principles, sophisticated instrumentation and applications in variety of fields. Recent advancements have helped in upgradation of these techniques in achieving high resolution of complex mixtures and reducing time and cost per sample. Hyphenation of these traditional techniques with spectral analytical techniques has further provided key advancements in separation along with analysis. UHPLC-MS is the hyphenation of ultra-high performance liquid chromatography with mass spectrometry offering on line separation and detection. HILIC involves the separation of polar analytes through their partitioning into a water-rich layer formed on the surface of the stationary phase. SFC has advantages over conventional liquid chromatographic techniques because of its low solvent consumption, excellent sensitivity, and solvent recovery. Capillary electrophoresis involved isolating ions based on their electrophoretic mobility in the presence of an electric field. HPTLC is an automated version of thin layer chromatography with advanced detection systems. 2D-LC is a powerful technique that combines two different separation modes to achieve higher separation efficiency and improved peak capacity. This book has also described the troubleshooting, problems and key challenges associated with these techniques.

Index

E

F

H

I

K

M

N

O

P

R

S

T

U

V

About the Editors

Dr. Rohit Bhatia, PhD

Dr. Rohit Bhatia is currently working as Associate Professor in the department of Pharmaceutical Chemistry at ISF College of Pharmacy, Moga, Punjab. He completed his PhD in Pharmaceutical Sciences from MRSPTU, Bathinda. He has specialization in Medicinal Chemistry, Computer aided Drug Designing and Bioanalysis. His area of interest includes anticancer drug discovery and paper based microfluidics. Dr. Bhatia published more than 65 review and research articles in various international and national journals. He has published more than 5 book chapters. He delivered more than 8 guest lectures as a speaker on various topics. He has qualified for the GPAT exam four times. Dr. Bhatia has more than 9 years of teaching experience. He guided 18 postgraduate students. He is a potential reviewer in various international reputed journals. He was the brand ambassador of Bentham science during 2019. He participated in and presented his research work at various national and international conferences. Dr. Bhatia has been granted with two patents by government of India. Recently he got a grant of 20 lac by MSME, India for the development of point of care device. He has also authored two books.

Dr. Bhupinder Kumar, PhD

Dr. Bhupinder Kumar is currently working as Assistant Professor at HNB Garhwal University (A Central University), Srinagar (Uttarakhand), India. He has completed M. Pharm (University Gold Medalist) and PhD in Pharmaceutical Chemistry from Central University of Punjab, Bathinda. During PhD, he was awarded with prestigious CSIR-SRF fellowship and Indo-Austria Scientist exchange visit fellowships. Dr. Bhupinder has got more than 55 papers published/accepted in highly acclaimed international journals and three patents (2 granted). Dr. Kumar has guided 12 MPharm student in research projects. He has attended number of workshops, conferences and FDPs. Dr. Kumar has expertise in medicinal chemistry (Drug designing, multi-step synthesis green synthesis, molecular modeling, screening drugs using various in vitro methods against enzymes and cell lines), organic chemistry, and green chemistry. His areas of interest include computer aided drug design (CADD) and synthesizing heterocyclic compounds of medicinal importance especially in neurological disorders, cancer and inflammation.